TWO
Awesome
HOURS

TWO
Awesome
HOURS

Science-Based Strategies to Harness
Your Best Time and Get Your
Most Important Work Done

Josh Davis, Ph.D.

HarperOne
An Imprint of HarperCollinsPublishers

HarperOne

TWO AWESOME HOURS: *Science-Based Strategies to Harness Your Best Time and Get Your Most Important Work Done.* Copyright © 2015 by Josh Davis. All rights reserved. Printed in the United States of America. No part of this book may be used or reproduced in any manner whatsoever without written permission except in the case of brief quotations embodied in critical articles and reviews. For information, address HarperCollins Publishers, 195 Broadway, New York, NY 10007.

HarperCollins books may be purchased for educational, business, or sales promotional use. For information, please e-mail the Special Markets Department at SPsales@harpercollins.com.

HarperCollins website: http://www.harpercollins.com

HarperCollins®, 📖®, and HarperOne™ are trademarks of HarperCollins Publishers.

FIRST EDITION

Library of Congress Cataloging-in-Publication Data
Davis, Josh.
Two awesome hours : science-based strategies to harness your best time and get your most important work done / Josh Davis. — first edition.
pages cm
Includes index.
ISBN 978–0–06–232611–9
1. Time management. 2. Cognition. 3. Performance. I. Title.
HD69.T54D385   2015
650.1'1—dc23   2015001465

15 16 17 18 19 RRD(H) 10 9 8 7 6 5 4 3 2

To my wife, Daniela, my parents, Susan and Don,
and my brother, Kenny, for their limitless love, support,
and appreciation for who I am and what I can do.

CONTENTS

TWO
Awesome
HOURS

Be Awesomely Effective

*W*hether we love or hate our jobs, the amount of work most of us have to do each day has reached unsustainable levels. We start a typical workday anxious about how we will get it all done, who we might let down, and which important tasks we will sacrifice—again—so we can keep our heads above water.

As we grab our first cups of coffee, we check our e-mail inboxes on our handheld devices, scanning to see who has added a new task to our to-do list. The stress builds as we read e-mail after e-mail, each containing a request that we know can't be dealt with quickly. We mark these e-mails as unread and save them for . . . "later." We mentally add them to the piles of work left undone the night before (when we left our offices much too late). More e-mails to answer, more phone calls to return, more paperwork to fill out. And everything needs our immediate attention.

In fact, too many things need our attention before we can even get to the tasks that really matter—and too many things matter. We frequently work all day long—at the office and then at home, taking care of our families, cleaning up, paying bills—sometimes only stopping to sleep. There simply isn't enough time, but so much always needs to be done.

If this sounds familiar, you are not alone. In my work as a professor, teacher, executive coach, author, and trainer, I have found these experiences are all too common among professionals and nonprofessionals at every level. More problematic, I see people from all walks of life—from executives to doctors to students to entrepreneurs to government workers—gravitate toward the same ill-fated solutions to find relief from this work overload.

And these solutions are making the problem worse.

Over and over, I watch smart, dedicated, hard-working people fall for the trap of "efficiency": We try to stay on task as much as possible, capturing any downtime throughout the day and putting it to use. If we have staff working under us, we try to get those people to do the same, for as many hours as they can pack into each day. Time-management books, gurus, and even whole consulting companies have taken up the challenge of helping us "do more in less time." As my brother, a Fortune 1000 executive, says, "We all have a lot of sh-t to do. It's good sh-t, but at the end of a day cramming ten pounds of sh-t into a five pound bag, we're still covered in sh-t."

Other time-management experts advise us to get to what

matters most first, because there may not be enough time for those tasks later. Yes, it's valuable to separate the truly important from the urgent though less important. But there's something frustrating about this advice. When all's said and done, there are still a lot of things we have to do that are not the most important on the list. Some things matter because they affect our relationships, some because not doing them will—in the long run—get us fired, some because we agreed to a deadline and we can't flake just because something else important is on that list.

Even if these tasks are not what matter most, we may still go home feeling anxious when we don't complete them. Sure, some problems go away if we ignore them. We can get better at letting things go. But with many of our tasks, we will not be absolved of responsibility for them. Eventually, they need to get done.

If a lack of efficiency were truly the problem, most of us, including my accomplished clients, would have solved our problems by now. Choosing the right system or app to help us manage our time or prioritize our tasks would relieve the pressure of the daily grind. But quantity and capacity are not the only factors that matter, and despite working as efficiently as possible, we are still not satisfied. Many of us feel stifled rather than accomplished at the end of a typical day.

However, what we want *is* attainable. Most successful people I meet want two things. The first is that they want to stop feeling out of control. The second is that they want to kick butt at work—they want to be masters of their craft.

In both cases, the typical and misguided response—which tends to fail—has been to expect ourselves to work constantly and pack more into our already packed days. What's wrong with this approach to getting work done? And how can we improve it?

A STORY OF PRODUCTIVITY

There's probably no one more famous for his industriousness than Benjamin Franklin. People the world over agree he was a model of effectiveness and productivity. He was frustratingly capable. His list of accomplishments is absurd: author, inventor, scientist, printer, philosopher, politician, postmaster, diplomat, and more. How can any human being do this much in a lifetime? A quick look at his rise as a printer and publisher—his primary profession—sheds some light on the way he worked and, in the process, reveals a lot about what we are doing right and what we are not.

By 1724, at the age of eighteen, Ben Franklin had already apprenticed in a printing house in Boston, worked independently in a printing house in Philadelphia, and published a handful of widely read articles.[1] That year he left for England, where he would learn the printing trade from the best, such as Samuel Palmer, a well-established printer. Not bad for a poor kid with sixteen siblings.

While working at Palmer's, Franklin quickly annoyed and impressed those around him with his work ethic and cleverness. His coworkers drank beer from morning to night; he drank water so he could have the physical stamina to outperform them and save a little money. You might say it was easier to have a competitive advantage in those days, but Franklin gets credit for seeing the opportunity, taking the risk, and following through. Ultimately, he was promoted and he moved to an even better firm.

When he returned to Philadelphia a couple of years later, he was willing to do what it took to establish himself. After working for another printer for a few years, he took on debt to set up his own business. With a print shop at his disposal, and in need of cash, he identified another opportunity: publishing his own material. There was only one newspaper in town, which Franklin considered "a paltry thing, wretchedly manag'd, no way entertaining." He knew he was the only printer in the area who also had the ability to write well, so he tried his hand at publishing newspapers and eventually *Poor Richard's Almanack*. Almanacs have space to fill, apart from their noteworthy dates. Franklin filled the empty spaces with his (now famous) proverbs, making his almanac more entertaining and much easier to sell. *Poor Richard's Almanack* was a hit.

In order to secure the success of his printing business, he also took on the position of clerk of the General Assembly, which allowed him to meet plenty of people who had a say in where government printing (things like ballots and money) was done, and he eventually landed the job of postmaster in Philadelphia, which helped him circulate his newspaper. These positions offered small pay and meant extra work, but they also allowed his printing business to take off, helping him become a man of some status in town.

Benjamin Franklin was and still remains a beautiful example of productivity and achievement. Work hard, take on more and more, and success will follow. Today, everyone thinks they have to be like Franklin to achieve some success. They have to do more than what seems possible. But the truth is, not even Franklin was like Franklin. As it turns out, beyond taking care of his finances, he was anything but focused on work.

We seldom talk about this other Franklin, hardly the live-for-your-job icon we sometimes think of. But I didn't have to look hard to find out more about him: it's in his autobiography. He loved to think and create. He spent huge amounts of time on hobbies and with friends when he could have been working at his moneymaking career as a printer. In fact, the very interests that took him away from his primary profession led to so many of the wonderful things he's known for, like inventing the Franklin stove and the lightning rod.

To understand the secret to his success, I believe it's crucial to look at how he spent his *downtime* and just how much of it he had.

One of his main hobbies as a young man was hanging out every Friday with a group of guys who were seriously into books and talking about ideas. The group would agree on a topic to discuss at the next meeting, and each would read what he could on the subject so he could come back prepared to argue. Books, however, were hard to come by in Philadelphia back then; many needed to be ordered from England. Franklin's group realized it would be nice to keep all their books in one place so they could check one another's references easily—a concept that led eventually to the great and historic public library now called the Library Company of Philadelphia.

Franklin did not found the library when he was around age twenty-five to make money for his printing business, nor was it part of a government position he held. He simply put time into founding this library because he enjoyed talking about ideas, especially ideas that would lead to improving himself and the world around him. He loved literature and art. He even wrote some music for his wife.[2] And, famously, he was an incurable flirt, spending a great deal of time wrapped in that pursuit after his wife's death.[3] He was also the original American self-help junkie. He tried vegetarianism briefly because he'd read about it in a book—and loved all the money he saved. Plus, he poured tons of time and energy into developing a plan to practice his now famous thirteen virtues. Of those thirteen virtues, one jumps out as seemingly relevant for anyone trying to pack in as much work in a day as possible: the virtue of Order (i.e., being organized). Franklin claimed he never really got good at that one, writing in his autobiography, "In truth, I found myself incorrigible with respect to Order; and now I am grown old, and my memory bad, I feel very sensibly the want of it."

He earned a reputation for enjoying the many pleasures of life—from learning to socializing to flirting to creating. It seems dazzling that he could do so much work professionally and still enjoy so much hobby, leisure, and social time. So how did he do it?

Every day he created the mental and biological conditions for peak effectiveness, and in those periods of effectiveness, he accomplished extraordinary things. He did not cram tasks related to his printing business into every available hour. In fact, in a plan he drew up for how to spend his days he included time for a two-hour break for lunch and other things, time in the evening for "music or diversion, or conversation," and a full night's sleep. It was probably because he made time for pleasure, learning, creativity, entertainment, physical health, family, and social connection that he was so successful in his money-making work, rather than in spite of it.

Devoting all of his time to his printing business rather than his other interests would have been the most *efficient* use of his time. But imagine how little we would know of him had he done so, had he never reserved the mental space and energy for his many inventions, for his philanthropy, and perhaps even for his printing empire.

Which Benjamin Franklin do you want to be: the one who carved out time for his hobbies and social pastimes, jumping from interest to interest? Or the one who outperformed his competitors to become a productive, well-regarded, and wealthy businessman? These days, it seems there isn't enough time for both, so we must choose to either enjoy life or succeed. The good news is that this is a false choice. We feel pressured to choose when we mistakenly assume that productivity depends on finding enough hours in the day.

THE EFFICIENCY TRAP

While helping high-level executives and professionals become more effective, I've learned that regardless of how high up the ladder we are, we typically respond to being overwhelmed by work in two ways. One is to force ourselves to stay on task without breaks in order to make the most efficient use of our *days*. The other is to work more hours—and to ask anyone who works for us to do so too—to make the most efficient use of our *weeks*. Underlying both of these solutions is the belief that to manage our workload we should stop "wasting" time—we should be "efficient." This belief follows from a fundamental misunderstanding of how our brains work.

Staying on task without a break and working longer hours are wonderful solutions for a computer or a machine. Computers and machines don't get tired, so the quality of work is identical every time they are used. Using them more frequently will only lead to greater productivity and efficiency. But, of course, we're not computers or machines. We are biological creatures. Continually demanding one kind of work—and a consistent level of effectiveness—from our brains is like continually demanding the same speed from a runner under any circumstances—whether sprinting or competing in a marathon, or whether running with no sleep after fasting for a day, jogging after recovering from a hangover, or exercising after being fed and rested.

There are consequences of being biological creatures on how we think. A number of people in the scientific community call these consequences "embodied cognition."[4] Embodied cognition includes the many ways that having a body influences thought. The brain serves as part of the control mechanism for the rest of the body. Cognition—any kind of thinking—cannot be properly understood without referring to the body it serves.

What does this mean?

It means that how you move your body may greatly affect your thoughts. Sitting with your hands behind your head and your feet up on a table—a pretty common "power pose"—can increase your level of testosterone and decrease your cortisol, a hormone combination that can lead you to both feel powerful and act like a leader.[5] Your physical movements may affect your moods too and can color your interpretations of other people's thoughts or intentions. Research suggests that if, for example, you make a hostile gesture—like flipping the bird—while evaluating someone, you are more likely to see the other person as hostile, because the movement primes the idea of hostility.[6]

Or consider how you learn: you rely on memories. But you don't install memories in your brain the way you install a software program or download a file into a computer. Rather, memories are something you grow. It takes time for neurons to structurally change so that they can more easily reactivate one another in the future, which may help explain why cramming the night before a test is not as effective as learning the material over multiple days, if you want to retain it long-term.[7]

These are a few of the thousands of findings illustrating the ways in which, by virtue of being biological creatures, we are quite unlike computers or machines and therefore cannot achieve the level of efficiency they do. However, each of us has vast untapped potential as a human that computers and machines do not have. And trying to be efficient all the time will block us from harnessing it. If my aim were to do ten thousand push-ups, I'd have a really tough time doing them without a break. But I would have no problem if I did a small number at a time between other exercises and spread them out over multiple workout sessions. The brain is very much like a muscle in this respect. Set up the wrong conditions through constant work and we can accomplish little. Set up the right conditions and there is probably little we can't do.

What I've learned from working with highly productive and happy people, and from my study of neuroscience and psychology, is that to be truly productive, our best bet may just be to ditch efficiency and create, instead, the conditions for two awesome hours of *effectiveness* each day.

TWO AWESOME HOURS

The key to achieving fantastic levels of effectiveness is to work with our biology. We may all be capable of impressive feats of comprehension, motivation, emotional control, problem solving, creativity, and decision making when our biological systems are functioning optimally. But we can be terrible at those very same things when our biological systems are suboptimal. The amount of exercise and sleep we get and the food we eat can greatly influence these mental functions in the short term—even within hours. The mental functions we engage in just prior to tackling a task can also have a powerful effect on whether we accomplish that task.

Research findings from the fields of psychology and neuroscience are revealing a great deal about when and how we can set up periods of highly effective mental functioning. In this book, I'll share in detail five deceptively simple strategies that I have found are the most successful in helping busy people create the conditions for at least two hours of incredible productivity each day:

1. *Recognize your decision points.* Once you start a task, you run largely on autopilot, which makes it hard to change course. Maximize the power of those moments in between tasks—that's when you can choose what to take on next, and can therefore decide to tackle what matters most.

2. *Manage your mental energy.* Tasks that need a lot of self-control or focused attention can be depleting, and tasks that make you highly emotional can throw you off your game. Schedule tasks based on their processing demand and recovery time.

3. *Stop fighting distractions.* Learn to direct your attention. Your attention systems are designed to wander and refresh, not to focus indefinitely. Trying to fight that is like trying to fight the ocean tides. Understanding how your brain works will help you get back on track quickly and effectively when you get distracted.

4. *Leverage your mind–body connection.* Move your body and eat in ways that set you up for success *in the short term.* (You can eat and physically do whatever you want on your downtime.)

5. *Make your workspace work for you.* Learn what environmental factors help you be on top of your

game—and how to adjust your environment accordingly. Once you know what distracts you or what primes your brain to be in creating or risk-taking modes, you can adjust your environment for productivity.

These strategies, derived from neuroscience and psychology, may sound simple; some may even seem like common sense. But we rarely employ them. Understanding the science behind them helps us know what's worth acting on and how to do so within the constraints we have. We can all learn to deploy them regularly and consciously with powerful results.

There's nothing magical about two hours. I'm recommending two hours because I've found that length of time to be both attainable and sufficient for getting to enough of what matters each day. The specific number of hours is not critical. As you gain experience with these strategies, you can set up conditions for four hours or even just ten minutes of peak mental functioning, depending on what suits your needs that day.

Note that I'm not suggesting you identify two specific and consistent hours every day (say, from nine to eleven A.M.) when you will aim to be effective. If you are like most busy professionals, you don't always have control over when things need to get done. If you are a morning person and your boss asks you to give a presentation at the next staff meeting—in the middle of the afternoon—

you better be in top mental shape when you deliver it. These strategies can help you set up the conditions for peak mental effectiveness at any time in your workday.

While I believe that you can accomplish great things under the right conditions, I'm not suggesting you'll be able to get all your work done in just two awesome hours. I do think, however, that when you are mentally effective, you can accomplish whatever matters most to you at that moment, with pride in your work and inspiration to do more. The rest of the day you can devote to those tasks that don't require much strategic or creative thinking: slog through e-mails, fill out forms, collect reimbursements, manage schedules, pay bills, plan travel, return phone calls. You can more successfully decide what to let go of among those tasks, too, when you're thinking more effectively.

Working in tandem with our biology—setting up the conditions for a couple of hours of peak productivity—allows us not only to focus on the tasks that are most important to us and our success but also to restore some sanity and balance to our lives.

MASTERING YOUR PRODUCTIVITY

Everyone is capable of learning to be as effective as they want to be. And in the rest of the book, I'll show you how. I'll describe how the five strategies work, explain the science behind them, and share stories that show them in action. For each strategy, I'll also suggest a few steps you can take to put that strategy into practice, and make it easy to use. It is my hope that every time you pick up this book you'll discover new insights, increase your self-compassion, and continue to master the game of being mentally effective.

At some level, we all know from experience that we can be remarkably effective in short amounts of time when we treat ourselves right—and horribly ineffective when we don't. Once you understand the science behind what makes us truly productive, I hope you'll trust and build on what you already know about yourself, and start thinking about your day in terms of how and when to set yourself up for two awesome hours.

—

RECOGNIZE YOUR DECISION POINTS

*A*s part of his responsibilities as a scenario-planning consultant, Doug writes a monthly analysis report of the latest developments in the clean tech sector. It's a task he loves—it allows him to be creative and to dig deeply into a topic that fascinates him.

While working on a report one day, he finds himself in a trance—not really aware of anything besides the computer screen in front of him and the noise of his typing. The spell, however, is broken when his eyes happen to drift to the clock on his desk. It is late in the morning. A twinge of slight nausea surfaces as he reluctantly recognizes that he ought to stop writing (even though he wants to keep going) and instead answer an e-mail he received days earlier from a colleague who needs his input to prepare their department's budget.

With trepidation, Doug opens his e-mail inbox, determined to answer his colleague. But a dozen new requests confront him. An e-mail from a project coordinator catches his eye. She needs to know his availability for the next few months to schedule a meeting. He tackles that request first, since it seems fairly straightforward and easy. But just as he is about to start answering the e-mail about the budget—the one he really needs to answer and that is fairly complicated—his calendar alarm buzzes to remind him that a one-on-one meeting with his company's CEO is in fifteen minutes. He still needs to prepare a little for it.

The smart thing to do would be to close his e-mail, get his thoughts together for the meeting, and start walking to the CEO's office. But the desire to feel a sense of accomplishment is too enticing. He wants to get that e-mail about the budget off his plate *today*. He's been feeling terrible every day he makes his colleague wait for it.

So in a split second, without really thinking it through, he decides to address that e-mail. He has a vague sense that in the past he was sometimes able to complete a similarly complex task in just ten minutes. It was unusual, but it happened. And there were times when he had pulled himself together before a meeting with the CEO in only five minutes. Who knows what bizarre, contradictory neurochemical signals are influencing his fingers to click on the budget e-mail, but that's what his brain decides to do: attempt to answer the budget input request.

Ten minutes later, he has barely collected the information he needs to start drafting his reply. Four minutes after that, he sees his meeting with the CEO is about to start, screams silently at himself, and hurries out of his office in frustration. He leaves his work in disarray. It will be time-consuming to retrace the information he's collected up to that point, so he can compose his reply, but he can't be late to meet the CEO.

He shows up to the meeting with the CEO frazzled and unprepared. His performance for the morning leading up to the meeting wasn't what he had hoped for either. He had made some progress on his monthly clean tech report but did not complete the final section. After the meeting, he loses more time trying to figure out how he intended to finish that part. He had managed to only start that e-mail to his colleague about the budget—but still the e-mail is not sent. On the bright side, however, he had sent a relatively unimportant email to the project coordinator with his availability so she can schedule a meeting months from now.

Throughout the morning, Doug operated on automatic, going from task to task but not making strategic decisions about how to best use his time. As we're about to see, Doug—like the rest of us—can't easily snap out of operating on autopilot. The trick to doing so is to recognize those relatively rare moments when we have a decision point—in between tasks—and to seize them.

PERIODICALLY, WE BECOME CONSCIOUS

Most of the time, we function in automatic mode—we think, feel, and act following nonconscious routines. Nonconcious refers to anything the mind or brain does that is not conscious. I don't mean to say that our behavior is thoughtless; I mean it is simply well learned and well rehearsed, and thus requires little conscious monitoring.

Although in the introduction I made a case that we are not computers that can perform predictably and consistently, in one regard we are very much like computers: almost everything we do—from flossing our teeth to answering a day's worth of e-mails—we do by following neural routines, the human version of computer programs, which guide our thoughts, feelings, and behaviors. We engage in these routines to some degree automatically—with no conscious review or consideration as to whether it makes sense to do so. Like a computer program, once we start a neural routine, we run it until it is either completed or interrupted.

If you start flossing your teeth, it's likely that by the time you're done, you won't even realize the number of intricate steps you took to reach the point when your teeth felt clean. If you start checking your e-mail a few minutes after arriving at the office, you might not even realize that after you finish opening, reading, and answering your first e-mail, you reflexively move on to the next . . . and the next . . . and the next, perhaps until you are interrupted by a colleague who has come to grab you for lunch. Chances are that when you got to the office that morning, you had hoped to tackle other projects, but once you started answering e-mails, the neural routines started running and you couldn't stop until something snapped you out of it.

In his book *The Power of Habit*, *New York Times* reporter Charles Duhigg shows how very like automata we can be, perhaps more often than we care to realize.[1] He explains that often we slavishly respond when we are presented with the right cues. For example, suppose you need to pick up some groceries on the way home from work. While driving to the store, you might not give thought to every action, like how hard you need to press on the brakes or when to look around. You can do each of these motions easily while almost completely focused on something else, like trying hard to remember your grocery list. When you get out of the car in the grocery store parking lot, if you're like most people, you probably don't think about putting your keys into your pocket. Later, you may wonder whether you have your keys on you—which are almost always there—betraying just how automatically you do it.

Much of what we do each day is automatic and guided by habit, requiring little conscious awareness, and that's not a bad thing. As Duhigg explains, our habits are necessary mental energy savers. We need to relieve our conscious minds so we can solve new problems as they come up. Once we've solved the puzzle of how to ballroom dance, for example, we can do it by habit, and so be mentally freed to focus on a conversation while dancing instead. But try to talk when first learning to dance the tango, and it's a disaster—we need our conscious attention to focus on the steps. Imagine how little we'd accomplish if we had to focus consciously on every behavior—e.g., on where to place our feet for each step we take.

In fact, our days comprise a series of habitual neural routines, what we often call "tasks": getting up in the morning, getting dressed for work, commuting to work, turning on the computer, answering e-mails, grabbing lunch, attending a staff meeting, going for a run, making dinner, getting ready for bed. The problem is that we often jump from task to task without giving much thought to what makes sense to do next. We respond reflexively or follow our impulses, however misguided these are. The result is a tremendous amount of time and energy being wasted.

The first strategy for creating a couple of awesome hours of productivity is very simple. This strategy is to learn to recognize the few moments during each day when you have the opportunity and ability to choose how you spend your time. These moments are when a task ends or is interrupted—say, you are done with a phone call—and you must choose the next task you are going to engage in: should you answer an e-mail or prepare for a meeting?

In my experience, we tend to rush through these moments, or decision points, in order to get back to doing something that feels "productive." Hurrying through one decision point—in between tasks—might save five minutes. Starting on the wrong task may cost an hour. But the five minutes hurts more because we're so aware of every second, while during the lost hour we're mostly on autopilot, so it hurts less. Sadly, many people waste hours doing work that either isn't critical or cannot be properly accomplished in the time allotted.

Another challenge is that since we are on autopilot so often, there are not many times each day when we access the conscious resources to make decisions about what to work on. So it's important to recognize these decision points and seize them. In the pages that follow, I'll show you how. But first, it will be helpful to understand how neural routines work and why decision points are so easy to mishandle.

ANATOMY OF A DECISION POINT

An influential theory in research circles is that we are "cognitive misers" in many ways.[2] Everything being equal, we tend to take the path of least mental resistance. Our nonconscious, well-learned neural routines are relatively easy for us to engage in. But our more deliberative, conscious decision making is more mentally taxing. So acting like cognitive misers, we are likely to rely on automatic neural routines instead of conscious decision making when we can get away with it.

We all get lost in a trance when we are going through the motions of a neural routine. Merriam-Webster's dictionary describes a trance in one of its definitions as "a state in which you are not aware of what is happening around you because you are thinking of something else."[3] If you are preparing for a presentation, you may be unaware of the two colleagues standing a short distance from your cubicle. If you are engrossed in reading a report, you might not notice that you are hungry or that it is time for lunch. While the neural routine is running, there is less self-awareness and less awareness of what's happening outside the routine.

But when the routine ends (e.g., when you finish flossing your teeth or reading the report) or is interrupted by someone or something (e.g., when a colleague interrupts your presentation preparations to ask your advice on another project), self-awareness ramps up. The transition from being deeply engaged in the neural routine to the routine stopping can be jarring.

Consider the example of a pediatrician, who conducts a couple dozen exams per day. In and out she goes from one exam room to the next, mostly following the same series of actions and behaviors: she says hello, she washes her hands, she approaches the child and does a physical exam while she asks the caretaker some questions, she takes her exam gloves off, she types information into a patient file on the computer as she continues to talk to the caretaker, she plays with the child for a few seconds, and she hands the caretaker a printed summary of the visit. While she is paying close attention to her patient and to the information the caretaker is giving her, many of her motions are fairly automatic.

But what happens if, when she finishes the exam, she finds the next appointment has been cancelled? She may feel a delightful moment of freedom, because she thinks of going down to the nurses' station to socialize with her staff. Or she may feel dread, because she would have no excuse not to get to the insurance paperwork that needs her attention. Regardless, she would find herself captured in a moment of heightened self-awareness and possibly indecision, as compared to when her patient-examination routines were running. When they were, she didn't have a decision to make about how to spend the next fifteen or twenty minutes; she simply moved on to the next patient.

Like the pediatrician, we frequently find ourselves adrift in these moments when a neural routine stops or is interrupted. But why?

To answer that question, we first need to understand two broad classes of mental function we engage in: self-conscious and deliberate versus automatic and non-conscious. A group from San Francisco State University believes that the main function of consciousness is to make decisions when our automatic neural routines run into problems—in particular, when different simultaneous neural routines guide us to do competing *physical* actions.[4]

For example, facing the computer while reading an e-mail competes with turning to face your wife while she's telling you about her plans for hanging out with friends (something that has *never* happened to me). These are two competing routines. One keeps you absorbed in zombie-like fashion in some imagined conversation with the e-mail recipient; the other requires you respond to and engage with your significant other in a live conversation.

These two behaviors—staring at the computer and turning to face our significant others—are incompatible, so the conscious ability to assess and make a decision comes to the rescue to help us resolve this conflict. When we detect a conflict that needs attention, a specific part of our brains—the dorsal anterior cingulate cortex—becomes active.[5] It is thought by some to serve as an alarm system meant to bring more of our conscious resources online.[6] Conscious reflection really only seems to operate as a stopgap, when our more automatic processing leads to conflicting actions and a need to make a decision.

Decision points, thus, often arise as a result of conflict—conflict between automatic behaviors, or between behaviors and goals. We may find ourselves pulled in many directions in these moments.

Because decision points often arise from conflict, they

can be unpleasant. In the previous example, finishing your thought while writing an e-mail and turning to face your significant other to hear what she or he has to say may be two tasks that you separately enjoy, but if you have ever had to choose between them, I'd wager you became irritated once or twice and found that decision point unpleasant.

In those moments when we become more self-consciously aware, we start to notice all kinds of things, like all the other to-do items we have forgotten, as well as the passage of time. Effortful control of what we're doing can feel like a drag. One study showed that the more we have to regulate our thoughts, feelings, and actions, the slower time seems to go.[7] However, just because we are more aware of time passing without being "productive" does not mean that a lot of time is passing. It just means we happen to be more aware of it. Most people I know, when they have a lot on their plate, get kind of anxious or feel guilty when they become aware of time passing without their making progress. Precisely because our decision points can be uncomfortable, we have a tendency to try to get them over with quickly.

And that's where things typically go wrong.

THE DOWNSIDE OF DECISION POINTS

The moment after a neural routine stops is one of the keys to two hours of awesome productivity. It's in that moment that you get to decide how your next chunk of time will be best used. Is this the best time to check your e-mail messages and answer as many as you can before your next appointment? Or if you have two hours free, does it make more sense to immerse yourself in a project that requires you to focus for a long block of time? What would be more effective: to prepare now for a meeting that will begin in two hours or prepare closer to the meeting time so the issues you want to discuss are fresh in your mind? Being intentional about what you plan to do immediately after you finish a task makes all the difference in the world in terms of how well you use whatever amount of time you have in front of you.

Being intentional about how we use our time is not an earth-shattering idea. Yet most of us rarely do it. Feelings like guilt, anxiety, or their positive counterparts—a desire to please others, eagerness—motivate us to get going on a task that the emotion relates to. But that task is not necessarily a good use of our time. The result is that when we become conscious of the clock ticking, fall for emotional urges to take on a specific task, or are overwhelmed by an increasing sense of indecision, it is only natural that we should latch on to the first task that comes to mind in that moment and get to it. We instinctively shorten the period of time when we are most aware that we are *not* working on a task—and we try to get started on another as soon as we can.

Sometimes we luck out and the task that happened to grab our attention turns out to be exactly what we most needed to do in that moment. But if that were a reliable strategy, then those feelings of regret we all know so well after "wasting" a whole afternoon on (you fill in the blank) would not be an issue.

Decision points can last seconds or often minutes. Imagine if you indulged every time you faced a decision point, taking around five minutes to transition from one task to the next. Maybe there are as many as ten decision points in a good day. So in total, you would have spent about fifty minutes engaged in decision points that day. Having taken the time to make a conscious decision about what to engage in, you've moved on to tasks that are either important to you or suitable for the time you have available to spend on them.

By contrast, suppose you tried to be "efficient" and plow through these decision points, these moments of "unproductive" time. There's a decent chance you would move on to a task that is unimportant or unsuitable for that moment. How much time is lost when that happens? The potential amount is huge. That's where so many hours of lost productivity collect.

In an effort to move efficiently and quickly through these decision points, we may miss the opportunity to direct our efforts and energy in an intentional way. However, we can learn to react differently to these moments.

MAXIMIZE YOUR DECISION POINTS

The key to making the most out of our decision points is taking a mindful moment to reflect on what's actually important to us in that moment. I actually disagree with some time-management experts who tell us that we need to do deep soul-searching to know what's important to spend our time on. In the grand scheme of things, I believe it's not hard to know what's important to accomplish. Ask any of us while we are on vacation what we think is important in our work, and I think we can easily answer the question. We know what matters.

But confronted with the pressures of the day, when it seems we are constantly responding to urgent tasks—answering e-mails that arrive with exclamation marks next to them, indicating they are a high priority (typically someone else's high priority, and often not our own), putting out fires when mistakes are discovered, reacting to everyday problems that always crop up—it's easy to forget what matters. We push what matters to the bottom of our to-do lists, because these items seem, as Stephen Covey refers to them in his book *The Seven Habits of Highly Effective People,* "important, but not urgent."[8]

For instance, in my case, anyone looking at me could easily say that writing books, papers, and blog posts is important to me. Learning how to listen well and ask helpful questions is important to me. Creating new syntheses of research and ways to teach about it is important

to me. Developing a team to help me do that is important to me. These are tasks that fulfill me, help others I
work with succeed, and advance my career. When I don't
spend enough time doing these things, they weigh on me
the most—more evidence of their importance. However,
what grabs my attention when I sit at the computer on
Monday morning? A thousand requests, promises to
others, and deadlines that have somehow found their way
into my consciousness. But before jumping into any of
these tasks, it is possible in just a few minutes to reconnect with what is important to me that day.[9]

When you reach a decision point—either when you
start your workday or when you complete a task and feel,
even for a split second, that sudden unpleasant moment
of confusion about what to do next—you have an opportunity on your hands. You can kick your self-conscious
processing mechanisms into high gear to make good
decisions about how to use this time on what matters to
you most.

There are three tricks to maximizing these decision
points:

- Savor each decision point.
- Plan your decision points in advance.
- Don't start a new task without consciously deciding
  it's the right one.

Let's take a look at each of these in more detail.

Savor Each Decision Point

A decision point doesn't come around that often in the day, and you can't always know when one will come. But this is the moment when you can willfully choose a new direction and, therefore, each one is precious. A decision point is to be savored; it is to be honored.

By honoring, I mean recognizing when you are experiencing one and seizing it. I mean allowing the decision point to happen rather than ignoring it and pushing ahead, on to the next task that your nonconscious processes nudge you toward. I mean taking a step back, reconnecting with what matters to you, and then deciding what the best next course of action is.

Decision points can give you some "distance" from your immediate concerns, allowing you to make more strategic and intentional choices. Research suggests that creating psychological distance leads to high-level thinking—keeping the big picture in mind.[10] When we are too close to a decision, we have a tendency to give too much weight to immediate concerns.

For example, research shows that people will forgo receiving more money a few months or a year down the road in favor of receiving substantially less money today, even though it's clear that what they will get right away is of much less value. The rate at which we tend to overlook the future value of things has even been plotted. One study showed that on average people would rather have a gift certificate for seventy-five dollars today than get approximately one hundred and eighteen dollars in three months or approximately one hundred and eighty-five dollars in one year.[11] Clearly, both one hundred and eighteen dollars and one hundred and eighty-five dollars are of more value than seventy-five dollars, but dangle it right in front of us, and it's so hard to resist.

While this example deals with greater psychological distances than the distance I'm suggesting we can get by stepping away from our work for a few minutes, it illustrates how unbalanced our decisions can be without some distance. Often we make decisions about things that will have consequences in the future. But when we get caught up in the moment, we lose track of the big picture. We're capable of taking an obvious logical decision and, by bringing it psychologically very close, overvaluing the present moment.

Let's go back to our friend, the pediatrician. When we left her, she had just learned that a patient had cancelled her appointment and there was no one in the waiting room to see her. She found herself with an unexpected twenty minutes of available time, an occurrence that happens maybe once or twice a week. Immediately her mind went to the pile of insurance paperwork that needed her attention. She also now remembers that she needs to complete some documentation for a medical student she supervised that morning.

At other times in her life, she would have sat at a computer and started cranking through either one of those tasks. She would have even told herself she was being efficient and that by doing this work now, she would be able to go home sooner. She would have undoubtedly gotten involved in a tedious and confusing form, the twenty minutes would have been gone in an instant, and she would have probably needed to drop what she was doing without finishing it, leaving her frustrated and hurrying to find her next appointment. Plus, she would have likely remained distracted for the first five minutes of that exam.

But in this instance, she knows better. She steps back from her racing thoughts and gets a little distance from the immediate concerns of the day, smiles a little and says to herself, *I've already set aside time for those things at the end of the day. It matters to me to be connected with the staff. I do better work, and I enjoy it more, when we all work together smoothly.* With that in mind, she heads to the nurses' station to chat with her staff. These women know how to have fun together—and our pediatrician knows that chatting with them will not only strengthen their bonds as colleagues but also give her brain a refresh before she returns to what will then be higher quality patient work.

Plan Your Decision Points in Advance

Interruptions and distractions are inevitable. Even with the most careful planning, our tasks are often disrupted by an urgent e-mail or phone call or by a colleague who pops by with "just a quick question." We may not know exactly when, but we know there's a high probability that these interruptions will happen. Every one of them creates a decision point.

Why wait until they occur unexpectedly to plan how you are going to respond to them? When caught off guard by an interruption, you are more likely to react rather than savor the decision point it has created. You are more likely to let your nonconscious mind quickly guide what task to tackle next—rather than taking the time to make a conscious, strategic choice—and that could mean wasting tons of time.

Planning the reactions we will have to our decision points before they happen allows us to maximize them and our time. A rather large body of research has shown that planning ahead for likely obstacles dramatically increases our chances of behaving how we would like to instead of just reacting. Planning in advance how we will behave in various situations has been shown to help people lose weight,[12] control their emotions,[13] and eat more fruits and vegetables,[14] among many other benefits. These plans are called "implementation intentions."[15]

An implementation intention is a plan to implement a certain action if a relevant cue arises. It is an if-then approach: *If an interruption occurs, then I will take X action.* You choose what that action shall be. Note that what does not work is to "plan" to use willpower in the moment and fight an urge, or to promise yourself that you won't resort to old, unwanted behaviors. Planning *not* to do something tends to fail. It's important to plan to take a new action in the moment, which you would prefer to take. Just planning a new action makes it more likely to succeed.

There is also mounting evidence that we use much of the same neural circuitry to visualize an action as to actually engage in it.[16] The evidence implies that visualizing the action we plan to take if an obstacle comes up preps the neural circuitry, and therefore makes us more likely to follow through on our plans. This may help to explain why visualizing successful sports moves (in addition to physical practice) has been shown to help people succeed at sports,[17] why visualizing surgeries can help doctors in training improve their skills,[18] why visualizing ourselves performing well in an upcoming job interview can improve our performance,[19] and why visualizing muscle activation can build muscle strength.[20]

The concept of implementation intention can be applied to creating more decision points too. That is, we can imagine events that are likely to occur during our day and form a plan to pause for a decision point when each of those events occurs. For example, imagine that several times a week you are interrupted by colleagues who stop by to ask a "quick question." You can decide ahead of time that when you are interrupted, you will turn that moment into a decision point. The interruption brings you out of your trance, out of the neural

routine that was guiding whatever task you were engaged in, allowing you to reassess whether you should
switch tasks. I sometimes get deep into a task that isn't
worth my time, and an interruption from the outside
can give me a chance to reevaluate how to use my time
more wisely.

By planning ahead and imagining new concrete reactions, we can build in decision points. Imagining if-then
action plans can maximize these moments.

As another example, let's rejoin Doug, the scenario-
planning consultant from earlier, as he sits down for
dinner with his family on a Sunday evening. Everyone
has just come to the table and the conversation has barely
begun when an insight about a work project hits him.
He is tempted to leave the dinner table to jot down his
insights and to spend some time noodling with the new
idea for his project. But if he does that, he will completely
miss spending the evening with his three- and six-year-
old kids, a loss that is regrettably frequent at this point
in his career. Still, he reminds himself, he needs to figure
out this project soon, and when he is at the office there
are always people asking for his time, e-mails to respond
to, and the usual list of challenges.

Although it's hard to plan ahead for a eureka moment
such as this one, Doug has noticed that for him, work-
related insights strike when he finally relaxes, typically at
dinner with his family. Thus, he has planned ahead for
this exact scenario: if an insight strikes while he is with

his family, he will take a couple of minutes to assess the importance of that insight.

On this particular Sunday night, when the insight hits him, interrupting the dinner conversation, a decision point is born. He steps back from his thoughts and recalls the broader situation. He is at home with his kids and wants to enjoy his time with them. But he would be crazy not to jot down the important insights he just got and risk losing them entirely. He decides to ask his family for a half hour to go get some thoughts down on paper and for the kids to come get him when the half hour is up. He knows thirty minutes will not be enough time to complete his work, so he commits to getting the ideas down in such a way that he can follow them up at work. After the thirty minutes are up, he heads back to the living room with his kids to build a truly killer Lego fort.

Don't Start a New Task Without Consciously Deciding It's the Right One

When you hit a decision point, quickly starting on the next task often leads to more wasted time than taking a few minutes to decide properly what that next task should be. The key is to seize the decision point moment. Here's how:

First, as soon as you finish a task, rather than think-ing about what you can do easily right away, *label this moment as a decision point*. For example, when I hang up the phone after a forty-five-minute coaching session,

I literally say to myself, "This is a decision point." That's enough to trigger me to pause. I sometimes even stand up and walk away from my computer or drink some water or coffee. Once I've let the mental dust settle, after running at full cognitive speed for forty-five minutes, I am more capable of deciding what's worth my time to start on next.

How do you decide what's worth your time? One key source of information is whatever is most important to you that day, which is much easier to recall after giving yourself a moment to step back and think. However, you can actually do much better than that, by also factoring in your current state, how tired you are, what mental resources you'll want at certain points later in the day, and in what environment you'll be doing your work. In the coming strategies, we'll explore ways to make best use of each of these factors to maximize every decision point.

USE YOUR DECISION POINTS WISELY

Our tendency to get lost in the trance of work, guided by our neural routines, is not a defect. Rather, it is a natural consequence of the fact that much of what we do is handled by brain processes that operate relatively nonconsciously and automatically. But our ability to set up two awesome hours of productivity depends, in large part, on making good, conscious decisions about how to spend our time and on what. Those rare moments in between tasks when we snap out of our work trance are a gift. We can take advantage of them thoughtfully.

As you become more aware of your decision points throughout the day, you may be surprised by how frequently you have in the past given in to your brain's desire to focus on whatever task it happened to come across next. Using your decision points wisely is the first strategy—and challenge. How to manage your mental energy when you work is the next one.

MANAGE YOUR MENTAL ENERGY

*E*very day is a battle of priorities. Should we handle the seemingly urgent request our colleague called us about last night? Should we respond to that new e-mail from our top client? Or should we work on that big report due in a few days? For years, productivity experts have said that the best way to manage our time is to focus on our biggest priorities first because there may not be time later. While they are partly right—it *is* often helpful to get to the biggest priorities first—their advice misses an important element. Our mental energy is the fuel that drives us—or fails to drive us.

Every task takes a mental toll on us; some even fatigue our minds. And perhaps every task elicits emotions that make that task and the ones that follow either harder or easier to do. While it would be nice to bring our A game to every task we tackle, there is only so much of the right mental energy to go around. In order to create a couple of awesome hours of productivity, we're much better off choosing what's worth giving the right mental energy to and putting off, in strategic ways, those tasks on our to-do lists that get in the way.

How can we successfully set up the conditions to produce the right mental energy at the right time? Once we understand what types of tasks tire our minds most and what role emotions play in our productivity, we can carve out at least two awesome hours a day when our priorities are our brain's priorities as well.

UNDERSTANDING MENTAL FATIGUE

Tom, a marketing director for a sporting goods company, is super excited about his latest idea: reviving the classic tennis line that had launched the company back in the day. The night before he is to give an informal pitch of his idea to the CEO, the CFO, and the rest of the C-suite, he lies in bed thinking about a dozen ideas for the line, fantasizing about how much the top leadership will love them and imagining himself as the guy who will be known for bringing back the company. Finally, he gets to sleep, much later than he'd hoped but totally pumped up and feeling good about the next day.

In the morning, while commuting to work, Tom ruminates on how he should organize his morning. His meeting with the CEO will be at eleven A.M. and he wants to spend at least thirty minutes jotting down some of the new ideas he came up with the night before. He is also aware that the last time he checked his e-mail was two P.M. the previous day, before he left the office early for an off-site meeting. He is anxious to find out if there is anything urgent that needs his immediate attention.

As the coffee finally kicks in, he sits down at his desk and instinctively opens his e-mail inbox. He thinks, *Let me just get through this stuff quickly so I can be caught up. Then I'll be free to prepare more for my pitch without worrying about having missed any important e-mails.*

An hour and a half later, he finally answers the last of the e-mails, opens a document to jot down some ideas for his pitch, and . . . sits there. He can hardly remember his great ideas from the night before. Of those he can remember, he has a hard time deciding which to work on. He will only have ten minutes to convince the company's decision makers. Should he lead with the customer feedback info he collected? Should he open with a survey of the competition? He worries that every idea is stupid or sounds naïve. All of a sudden, he isn't so sure about this new project. *Do I even have the chops to pull this project off?* he asks himself.

The truth is that in that moment, Tom does not have the ability to pull off this project or the pitch. His brain's executive functions are exhausted—and he doesn't even know it.

Why Saying No to a Donut Tires Your Brain

Without knowing it, Tom overtaxed what psychologists and neuroscientists call his "executive functions," which is a term to describe the various control and direction-setting tasks the brain handles. Parts of the brain act much like the executives in a company act in relation to subordinates—directing and course-correcting their behaviors—but in the case of the brain, it is in relation to our thoughts, feelings, and actions.

The executive functions that the brain handles include decision making (*Should I wear a red or a blue shirt this morning?*), planning (*First I'll go to the dentist, then on the way from the dentist to my house, I'll stop to pick up dinner*), and holding on to thoughts for a short time while we need them (*I need to remember the name of the person I just met long enough to introduce her to my business partner*).

Executive functions also involve inhibiting some actions, feelings, or thoughts. For example, letting personal slights from the boss go or staying focused on putting together presentation slides amid distractions like phones ringing or e-mail notifications popping up. Self-inhibition, or self-control, is one of the major executive functions the mind performs.

Engaging in self-control tends to wear out our self-control. Various forms of self-control—such as watching what we eat, controlling our anger, limiting impulse buys, avoiding the urge to sell stocks when prices have fallen—

have at least one thing in common. After we engage in them, we typically act as though our brain mechanisms for self-control have been depleted or fatigued. Interestingly, brain research on the topic suggests that the brain is still capable of self-control when we are "fatigued" in these ways, but we seem to lose the motivation to keep controlling ourselves.[1]

The jury is still out, but it has been proposed that self-control of any kind (resisting a pastry at breakfast, holding back tears, turning the other cheek when someone gets aggressive, and many other forms) rely in part on a few shared brain regions, such as the ventrolateral prefrontal cortex and the dorsal anterior cingulate cortex.[2] After completing a self-control task, those regions of the brain are less useful when tackling another self-control task. It's a lot like exercising physically: after going for a run, you've satisfied your need to work out, so it's not as compelling right in that moment to go for *another* run.[3]

Unlike going for a run, which most people tend not to do more than once a day, our brains are dealing with self-control all day long—helping us to avoid procrastinating when we have a project to finish, to say no to another slice of cake even though we really want it, or to overcome the urge to stay in bed when the baby wakes up crying in the middle of the night. As we deplete our self-control, it becomes that much harder for our brains to handle later self-control tasks, which explains why if we say no to eating a donut at breakfast, it may be harder to say no to a cupcake at our colleague's birthday lunch, and even harder to say no to dessert after dinner. Our worn-out brain finally caves in, and we find ourselves wolfing down half a gallon of Ben & Jerry's New York Super Fudge Chunk ice cream.[4]

The Consequences of Making Too Many Decisions

There are probably very few jobs for which chocolate ice cream avoidance is critical. But every professional, executive-level, or managerial job requires quality decision making and self-control in the face of many potential distractions. At its basic level, being productive requires that we stay focused on a task and resist the many distractions in a work environment: e-mail alerts, colleagues stopping by, other more enjoyable tasks waiting to be done, and so on.

At a deeper level, being productive requires self-control because quality decisions, investments, or plans require that we deal with competing options. Whenever there are competing options, we have reasons to pursue each option, and therefore self-control is needed to say no to all but one. As Roy Baumeister and John Tierney explain in their book *Willpower,* people keep their options open—sometimes even at great cost and little gain— because "it takes willpower to make decisions, and so the depleted state makes people look for ways to postpone or evade decisions."[5]

This link between decision making and self-control is

where things get especially interesting. Not only does it seem that as we deplete our self-control reserves for one task it makes it more difficult to motivate ourselves to do well on other ones; there's evidence that other executive tasks, like decision making, may take a toll on our self-control too.

What's more, research led by a group from the University of Minnesota and Florida State University has illustrated how easy it is to deplete the brain in these ways without even realizing it.[6] In one of their experiments, participants were asked to either choose which courses they might take for their requirements or simply look through the course catalog and think about the courses pertaining to their requirements. So the groups did essentially the same things—except one group had to make decisions, while the other group did not. Then the researchers examined how much mental energy the participants had left in the tank, by seeing how long they would persist at a task. All participants were invited (but not required) to study for fifteen minutes for an exam that the researchers would administer later.

To make this mimic real-life temptation for students and therefore the need for self-control, during this fifteen-minute period, all participants were free to play a video game or pick up magazines in the waiting area instead of studying like they knew they should. The participants who had made course decisions gave up studying after approximately just eight and a half minutes, while those who hadn't made decisions spent about eleven and a half minutes studying. In other experiments by that team, having participants make decisions—about video clips, test question styles, and other factors that would influence how a class they were in would be taught— decreased the participants' self-control afterward. The act of having made those decisions led students to spend only about nine minutes on unsolvable puzzles, while their colleagues, who had made no decisions immediately beforehand, gave the task about twelve and a half minutes.

In subsequent experiments, the same researchers showed that neither the size nor the importance of the decision, nor even the specific nature of the self-control task, seemed to change the toll that making decisions had on the participants' tendency to exert self-control and persist at a task. In other words, making even typical, everyday, unimportant decisions can leave someone less mentally motivated for self-control immediately afterward.

In short, decision making leads to mental fatigue, which reduces our ability to perform at our best.

And this is exactly what happened to our friend Tom, the marketing director who was fruitlessly trying to prepare for a new product pitch. What he did not realize is that answering e-mails, although it seems like an easy task, is actually quite taxing. Answering every e-mail involves making decisions, sometimes complex ones: *Should I reply? Do I have to respond now? If I write it this way instead of that way, will the person react well to my e-mail or be offended by it? Should I delete it or file it for future reference? Should I write a short response or forward it to someone else?* When answering e-mails, Tom has to make decisions about time, value, social consequences, alternative future scenarios, and perhaps emotional consequences. Even if an e-mail is, in the scheme of things, unimportant, there's still a good deal of decision making that happens.

Tom was right that if he could spend even thirty minutes jotting down his ideas before his meeting—just thirty *awesome* minutes—he would be in great shape to deliver his pitch to the company's higher-ups. But he was wrong to think it's just a matter of finding *any* thirty minutes to do it. He is faced with a project that needs creative decision making, persistence, and initiative. Unfortunately, he spent the prior hour and a half fatiguing his brain's executive functions by answering e-mails that weren't that important, depleting the very resources he needs to do well at the task that matters the most to him. If he were a race car driver, it would be like taking his race car through city traffic to get to the race—not the fresh start he'd want.

When he finally turns his attention to preparing for his pitch, the time he spends on task is worse than wasted. Since his brain is fatigued, the same decisions that would have taken him minutes to make when at his mental peak are hard to make at all. His self-control resources are spent, making him susceptible to giving up on his ideas, even to losing confidence with the project. In the end, he walks to his pitch meeting feeling uncertain about his ideas—hardly the energized feeling he dreamt about the night before.

UNDERSTANDING THE POWER OF OUR EMOTIONS

In the previous section, we saw how it can help to pay attention to mental fatigue when deciding what tasks to engage in, to ensure we are setting ourselves up for a block of truly productive time. But there's another way we can ensure that our brains are ready when we need to perform at our best. When we understand how the tasks on our to-do lists elicit emotions that make completing those very tasks, and the ones that follow, either harder or easier, we can plan ahead accordingly. Because being "on" at the right moment matters so much, saying no to tasks that will get in the way of that is key to deciding what should get our attention.

Although we may not always be aware of it, many of the tasks we perform—whether we're answering a colleague's e-mail or engaging in a tough negotiation with a supplier—elicit emotions: excitement, anger, pride, boredom, uncertainty, anxiety, and so on. The emotions can be mild or strong. And since emotions have a deep effect on how well we perform, knowing what to expect from our emotions can open up whole new opportunities for a couple of hours of awesome productivity.

How Emotions Prime Us to Perform

The reason emotions have a surprisingly powerful effect on our performance is that emotions have adaptive value— that is, they have the ability to help us cope with and respond to the situation at hand. For example, Beyoncé was once quoted as saying, "Every time I get on the stage, I'm nervous [beforehand]. I'm actually really scared when I'm *not* nervous." If she's not nervous, she explained, she does not perform at her best.[7] It's hard to imagine someone who looks so comfortable up on stage feeling nervous. But it is precisely that experience of anxiety that prepares her to perform at her best. Emotions—even some we may perceive as negative, such as anxiety—are great tools for getting our minds energized and focused for the task ahead.

The idea that even negative emotions may help us adapt to a situation we're faced with runs directly in the face of some common beliefs. Imagine the star quarterback of a football team sitting in the locker room at halftime. He has just let his team down in the most important game of the season, throwing a terrible pass that the other team intercepted to run in for a touchdown. After his confidence was shattered, he couldn't orchestrate a single successful drive. Clearly, his emotions were getting in his way.

Massively frustrated, he stands up and starts pacing the locker room, until his frustration gets the best of him and he throws his helmet down violently. Still, his emotions are not helping much. However, finding his opportunity, the quarterback's coach encourages the young man: "Go ahead, get angry!" After a two-minute inspirational speech, the quarterback is fired up, ready to return to the field to do what he has to. "Out of my way, Coach," he says. "I'm gonna go out there and kick some butt!" The coach has helped him turn his frustration into anger, and anger has primed him to go out onto the field laser-focused on what he's moving toward, ready to perform. Anger can be quite adaptive in the context of a football game.

Being angry or getting upset prepares us for certain kinds of thoughts and actions. With each of the following common emotions, we'll see the various ways in which it primes us for particular behaviors or tasks. Both positive and negative emotions have utility. But let's start with the negatives, since that is more surprising for most people.

Anger
Anger is unusual amongst negative emotions in that it facilitates approach-oriented behavior.[8] Approach-oriented behavior refers to actions that move us toward a person, object, or idea.

Typically, when you approach positive things, anger plays no part. For example, if you think about how good a dark chocolate truffle would be, you may have the urge to go find one and consume it. In that case, you are orienting your behavior toward seeking out the chocolate.

Sometimes, however, we will approach something even though it won't necessarily be a pleasant experience. That's when anger can be handy. For example, a store owner considers increasing prices on the items she sells in order to increase profits, which she badly wants. But by doing so, she will run the risk of violating the trust she's built with her customers. She is afraid that increasing prices will lead to customer backlash, which she wants to avoid. Tapping into a source of anger, rather than fear, may help her take the step toward achieving what she wants—increased profits—even though there may be an unpleasant period as her customers adjust.

So how does she get angry at will? When I coach someone on emotion regulation, they more often want to get better at controlling their anger, not eliciting it! But there's a time and a place for everything—even negative emotions. One option is for the store owner to recognize when she's angry for some other reason and use that as a good moment to consider pricing options. Another is to reframe her situation. For example, she may ruminate on how unfair it is that her profit potential is held hostage to the whims of shoppers who don't understand the value of the products she sells.

The next time you're afraid to take a risk even though you know it's the right thing to do, consider getting a little angry. As a mentor of mine once explained, sometimes the most motivating thing a person can hear to finally confront a challenge before them is a pronouncement by some authority that they will never be able to overcome that challenge. With angry indignation, they instantly set out to prove that authority wrong. (Of course I don't encourage people in positions of authority to say such things to those they manage or oversee.)

Sadness

Sadness has several surprising effects. When we're feeling sad we tend to be less biased in decision making—thinking a little more slowly and deliberately about whom to trust, for example. We are also likely to act more fairly and less selfishly. Plus, we err on the side of healthy skepticism, helping us to avoid being too gullible. And we're more apt to put in the effort to make a message persuasive. All in all, it seems that when we need to slow down, be thoughtful, and think critically, sadness can be a real resource.[9]

So if you're on the receiving end of a sales pitch, you may want to remember how much you miss your childhood dog. And conversely, if you're feeling super happy, you may avoid taking a sales pitch and use that positive emotion for something it's good for instead—like getting to your creative work.[10]

Anxiety

My father is a psychiatrist. When I was a teenager, I was very fortunate to be able to talk with him about feeling nervous about some upcoming big challenge. He explained to me that anxiety and readiness are physiologically almost the same. Anxiety is a way of revving us up to be on high alert and ready to react to whatever may come our way.

There are certainly times when you need to be highly alert and ready to react—for example, in a presentation, during a meeting you're running, or during a sales call you need to nail. If you don't know what anxiety can do for you in a positive way, then it's easy to just wish the feeling away; but when you understand its benefits, you can be grateful for it.

The next time you find yourself feeling nervous, see how it goes to say to yourself, *No, I'm not nervous. I'm alert and ready to react.*

Scientific evidence also suggests that on some level we have figured out the value of anxiety—at least our behavior makes it look that way. One set of studies showed that some people prefer to feel anxious when they have a challenging task coming, and they tend to perform better when anxious.[11] It seems Beyoncé may be onto something by wanting to be a little nervous before going on stage.

———

ALL THIS TALK about the value of negative emotions may make it seem as if I'm encouraging you to feel bad. Mostly, I am not. As you probably would expect, positive emotions also have an important impact on our performance.

Feeling good is especially useful for discovering new insights,[12] being creative,[13] being less critical when making decisions,[14] and making snap decisions.[15] The feelings themselves, by virtue of being experienced as pleasant, are in part their own reward too. And positive emotions can lead to easier collaboration, presumably through the positive expectations that they generally provide.[16] The research is not always clear on specifically which positive emotions have what effects.[17] But I believe it is safe to say that positive emotions—happiness, joy, amusement, or generally feeling good—have these effects on our performance.

If you want to be able to let unimportant things go more easily—look with a less critical eye some of the time—a positive mood can help.[18] If it's time to get creative, I suggest first getting into a positive emotional state. When you're going to need to make decisions fast and there just won't be time to deliberate, see if you can enter that situation in a positive mood.

One way to do this is to notice whether you're in a good mood when you come to a decision point. Another way is to influence your emotions. Try this: Close your eyes and remember something that makes you generally happy—a favorite TV show, learning, chatting with a friend, relaxing for a few minutes with a book, getting some exercise, eating when you're hungry, fantasizing about something you want, thinking about sex, having a good laugh. Remembering something that is emotionally positive helps to bring about positive emotion. Obviously, engaging in those things that make you happy, as opposed to simply remembering them, should help too!

HOW TO MANAGE YOUR MENTAL ENERGY

Now that you have a deeper understanding of how our brains get fatigued and how our emotions can drive us, we can apply that knowledge to set up a couple of awesome hours of productivity. The consequences of our mental energy on our abilities are real and biological.

Here are a few ways to manage your mental energy when you know that you have to be at your best.

Limit Your Mental Fatigue

Most tasks, at least for professionals and knowledge workers, lead to some mental fatigue. After all, we are constantly engaging in activities that involve decision making and self-control. The key to limiting mental fatigue is recognizing the work that is most likely to deplete your resources in a *substantial* way and, when you have any say in the matter, to simply not engage in that work *before* you want to be at your best.

For example, as a psychotherapist, Helen has to make a real effort to stay compassionate with her patients. Sometimes this is quite difficult for her, especially when she's dealing with patients who are caustic, stubbornly self-destructive, or involved in marital infidelities. But staying compassionate helps make possible the hard and necessary work of talking through these issues. However, Helen does not practice her profession in a vacuum. She has challenges in her own life too and multiple clients to see in a day.

Because remaining compassionate requires a good deal of self-control, Helen keeps an eye out for those mental tasks outside of her client sessions that really deplete her self-control resources. In the past, when one of her relatives was going through a difficult time and needed to talk to her, she made herself available at any point—and typically ended up exerting a good deal of mental effort to stay compassionate with that family member, depleting the resources she so needed to do her job well. So while she continues to help her relatives through tough times, she doesn't do it right before a patient's session. This simple shift has allowed her to maintain the exceptional client work for which she is highly regarded and still be a huge support to her family.

So how can you identify the tasks that lead to mental fatigue and keep you from being incredibly productive? If you feel spent after doing a task, there's a good chance it is tapping into your self-control. The degree to which tasks take a toll on self-control, decision making, or other executive functions varies with each person. For example, someone well practiced at proofreading will not need much of their executive functions to do it—they can do it mostly on autopilot. For someone not practiced in proofreading, who also has a hard time sitting still and tends to value the big picture over the details, proofreading will probably take a very large amount of self-control.

Here are some examples of common activities that can lead to mental fatigue:

- Switching frequently between one task and another
- Networking and making small talk
- Sitting still for hours
- Making cold calls
- Identifying errors and correcting them
- Planning or scheduling projects
- Keeping track of deadlines

Avoiding these activities may seem hardly practical, since the higher our positions at work, the more likely we are to have to make decisions, plan, and collaborate with others pretty much all the time. The important thing to remember, however, is that we don't have to completely avoid these activities. We don't need to be great all the time, and we can't be. We are not machines and cannot produce the same level of work every minute of the day. But if we strategically choose the *order* in which we complete the various tasks on our to-do lists, we can carve out two awesome hours when our brains are not as fatigued and get some amazing things done.

Do you hope to spend two hours mapping out a new initiative for your department? Don't tackle the project right after mapping out a different initiative. Do you need to write an e-mail to your boss outlining the reasons why a top-priority project is not on time and asking for more resources to complete it? Don't even think of starting that new e-mail after you've already been dealing with e-mails for an hour and a half. Does the human resources department want you to complete your employees' annual review by the end of the day? If it matters to you to get it right, it would be better not to do it at the end of the day, because the odds are you will have a good amount of mental fatigue by that point.

Even some of the common ways in which we pass the time when we are taking a break—presumably for the purpose of refreshing our minds—probably fatigue us even more and should be avoided if they occur just before we have to be on top of our game. For example, if you often turn on the news or check out a news website that reports on the latest tragedy or an upsetting political development, it can require a good deal of self-control to manage a knee-jerk reaction to these kinds of stories. It's also easy to get carried away emotionally by them (e.g., getting riled up over a political figure's latest scandal). So avoid these activities before you have to be at your best.

Here are four things to help you avoid mental fatigue that you can try this week:

1. Complete your most important work first thing in the morning, before your brain has been depleted by hundreds of small decisions. Think about the most creative and interesting task on your plate right now, or the one with the biggest long-term upside, and spend one to two hours first thing in the morning on it. And when I say first thing, I mean first thing—that is, before checking your e-mail or looking at any media, such as TV, newspapers, smartphones, or computers.

2. Consider the tasks on your to-do list for the day, and label each of them as "Important Decisions," "Creative," or "Other." Carve out time late in the day (perhaps after your lunch, during your food coma?) to complete the tasks in the "Other" category. Knowing you've scheduled time for these makes it less likely that you will try tackling them earlier in the day, when your mental reserves are highest.

3. Try reading and responding to your e-mail messages for only one hour in the afternoon and reflect on whether doing so improved your ability to focus more clearly on tasks that require problem solving or creativity during the rest of the day. I know this is a terrifying suggestion for some people. And it's true that some days don't allow for this, but try it once or twice and you might be surprised to find it more possible than you'd feared.

4. Make a few decisions the night before you have
a big day, so you won't have to make them on the
big day. They can be small (like what to wear or
have for breakfast and lunch) or they can be large
(like deciding what tasks actually matter to you to
accomplish on the big day). Organize your to-do
list based on those large decisions.

Here are three ways to refresh yourself if you get fatigued
or overly emotional and need a quick replenishment:

1. Breathe deeply and slowly for a bit. Breathing
helps you directly change your physiology, and
feelings are, in part, the experience of what's going
on in your physiology. So breathing can directly
change what you feel when you have feelings.[19] For
example, heart rate is strongly tied to breathing
rate. In general, the faster you breath, the faster
your heart will beat, and the slower you breath, the
slower it will beat.[20] So breathing slowly really can
help calm you down. This is one place common
sense has it right.

2. Have a good laugh. A positive mood has been
shown to help restore us when we are mentally
fatigued.[21]

3. Take a short nap—and I specifically mean short.
Researchers at Flinders University in Australia

found that a ten-minute nap helped reduce fatigue
and improve alertness as well as a variety of
cognitive tasks, with the benefits lasting for about
two and a half hours. And while a twenty- or thirty-
minute nap was also restorative, it was in some ways
less restorative than ten minutes, because it took
longer for people to be alert enough after the nap
to start benefitting from it, without extending the
benefits beyond two and a half hours.[22]

Anticipate the Emotions Certain Tasks Elicit

If some jerk cuts you off on the highway on the way to
work, will it make you angry? Does your husband or
wife always get irritated when you leave the dishes out?
Does your boss always get anxious before a big event and
stress out everyone on the team? Do you have a friend
who loves videos of cats doing cute things? We're capable
of picking up on people's emotional triggers. We can't
predict that the trigger will happen on a given day, but *if*
it does, there's a good chance we know what emotion is
coming. In this way, emotions are predictable. And if you
know what emotions to expect, then you can make sure
to schedule your two hours of productivity around them.

As you glance through your day's calendar, you probably have an idea about what may happen emotionally
in the day—and it probably doesn't require a lot of deep
introspection. For example, is there a presentation on

your schedule? If you are afraid of public speaking, then you can expect to feel anxiety. Is there a long meeting on the agenda that will keep you from fulfilling a deadline? Expect to feel aggravated or frustrated. Do you have a performance management conversation coming up that you are not looking forward to? Expect to feel dread and maybe insecurity. Whatever emotions you predict will follow these events or tasks may well remain with you for a couple of hours, depleting your mental energy.

On a more positive note, will you be working on a project with a friend next week? That social time can feel great and put you in a better mood. Plan on working on a task that requires creativity and problem solving after working with your friend, when your positive emotions are high.

Practice Strategic Incompetence

I have encouraged you thus far to set up a couple of awe-some hours for your most important work and save your unproductive time for everything else. But there are days when we'll have to choose not to do some things on our lists at all. Sometimes we may be better off letting certain things go in order to be ready when it matters most. It is better to spend two awesome hours on something with meaningful potential to bring in business or advance our careers in the long term than to have no awesome hours while trying to get to everything on the list.

As I see it, here's the biggest difficulty with that advice

for many people. There's a reason that everything is on our lists, and there's often another person who cares about whether each thing on the list gets done. As thoughtful and responsible people, we care about not letting others down, and we care about showing others we are competent. The toughest part is that it's not just an issue of prioritizing or knowing what matters most. It's that whenever there is a decision to make between priorities, there are always some valid reasons to try to do it all. And there are always some consequences to not doing it. But we can lose the battle to win the war.

When speaking with a then up-and-coming—now successful—corporate executive, I once jokingly called this approach "strategic incompetence," to capture the feeling of failure that he and I were both aware of when we let some things go. We each laughed at first, but putting that frame on the idea freed him up to view letting things go as a strategic choice, rather than a dilemma. This is not just about prioritizing, but about empowering ourselves to make choices some other people might not like, or that we'd rather not have to make.

Perhaps "incompetence" is too strong of a word, but I use it because of the all-too-common experience that when we consider letting something go it can feel that way—that we are being or looking incompetent. It really does take some discipline, especially at first. But we are human, so we have to work with the mental energy we can muster.

Choose your moments of peak performance and sacrifice liberally around them. To leverage your mental energy, do a few things that really matter excellently, rather than doing everything moderately.

There are some items on your agenda for which you will absolutely want to be at your best. And there are other things that you will do out of obligation, fear of getting in trouble, or other reasons besides their being your top priorities. Be careful about saying yes to everything on that list. Trying to do it all will crowd out the mental energy you need for what matters most. Moreover, other people will come to rely on you for those things. It can be a vicious cycle, making it harder to say no or ignore the requests in the future.

You can be at 50 percent of your best all day long— answering e-mails, keeping up with the latest news, attending every meeting you're asked to join, prepping presentations, and so on—or you can ignore one of those on a critical day and be at 100 percent of your best for the work that matters most. From what I've seen, after people experience the value firsthand of saving the right mental energy for the right work, they can find it easier to make such choices in the future.

Choosing your moments of peak performance can feel like a gamble, and it can be one of the tougher things to do that I mention in this book. But it is worth it, because the right mental energy for the right task can mean the difference between awesome productivity and a weak turnout.

MAXIMIZE YOUR EFFECTIVENESS

So let's start Tom's actual presentation day all over again and see how he might have better scheduled his tasks had he anticipated the effects of mental fatigue and emotions on his productivity.

Tom finally fell asleep the previous night after being super excited about the many brilliant ideas he'd come up with. Once in the office, he decides to check his e-mail. Since he hasn't checked it since two P.M. the previous day, he needs to make sure he isn't missing any important messages. He is aware, however, that he didn't get much sleep, so his mind is not as fresh as it could be and reading and responding to e-mails will involve a lot of little decisions that may exhaust whatever mental energy he has.

He agrees to give himself only thirty minutes to do a targeted search through his e-mails, just looking for anything unusually urgent, and he even sets an alarm on his calendar. He is bound to find at least one e-mail that will set him off emotionally, such as an e-mail about a fire he needs to put out or an e-mail from someone who usually irritates him. The emotions these e-mails may elicit will also affect his performance in his next task. So after thirty minutes, he stops. The urge to continue answering e-mails is huge—so many more e-mails have popped up since he sat down, even. But he reminds himself that he has to save his mental reserves for the most important task of the day: doing the final preparation for his pitch.

His brain is already getting fatigued, which he realizes only after he stands up and walks away from his
computer. He needs to clear his mind. So he first takes a
few deep breaths, then goes to grab a cup of coffee at his
favorite coffee shop a block away. He is only away from
the office for five or ten minutes, but the movement and
change of scenery really helps. He is now in great mental
shape for an hour of awesome productivity. He sits down
and straight away—without looking at any media—starts
reviewing all the ideas he came up with the night before,
decides which are best to present, and then devises a creative way to tie together all the pieces for what he is starting to feel will be a very powerful pitch.

An ordinary day for Tom as a marketing director involves many skills, like creative decision making (e.g.,
developing a new project proposal), emotion management
(e.g., staying calm while receiving or giving tough feedback), and analysis (e.g., understanding the new budget
constraints and considering potential consequences on
the team). Only Tom knows which of these are energiz-
ing for him, which are draining, and which have the
potential to lead to some emotional consequences. Only
Tom knows which are really important to him and which
are obligations of the job.

Like Tom, we can figure out what our priorities are
and recognize the social situations that elicit our strongest
emotions. It's time to use that information. Combined
with a deep understanding of how our brains become fa-

tigued and how our emotions can either derail us or help us, this knowledge is key to our couple of hours of awesome productivity.

The next time you find yourself at a decision point, free to start a new task, look back: What tasks have you just been engaged in? Have you exhausted your brain with too many decisions? What are you feeling right now and how is that likely to impact your next task? And the next time you consider your calendar, notice what is planned right before and after a task that matters the most. Have you scheduled your critical task when your mental energy is most likely to be drained? Or have you set yourself up for success? And when you need your mental energy for something critical, strategically let some other things go.

Managing mental energy gives a science-based approach to planning ahead and scheduling your tasks in a way that maximizes your effectiveness in whatever time you have.

STOP FIGHTING DISTRACTIONS

*Y*ou can now recognize the decision points in your day—lucky you! You may realize that by choosing wisely what task to take on next, based on the levels of your mental and emotional reserves, you are well on your way to a period of awesome productivity. The next step is to make that period as effective as possible once it is underway. In other words, now you need to focus for a prolonged period of time.

Staying focused is not easy. Let's take a look at a typical workday to see some of the common challenges.

By ten A.M. Amanda, a freelance web and user-experience architect, is frustrated. She spent all morning grinding out invoices for her customers, including one for a large client who is late in paying. Now, annoyed that she spent two hours doing paperwork, she turns her attention to the real priorities of the day: the deliverables she promised to three different companies.

Looking at the clock, she realizes that at least one will need to be postponed—there is not enough time to finish all three. Which two should she focus on and which one should she postpone? Annoyed at herself for still not having done anything productive that day, she randomly starts working on one of the projects; she has already pulled out the notes for it, so without really choosing, she figures she might as well get started with that one.

But she can't seem to let go of her irritation over the invoicing—a process she hates—and her anger toward the delinquent client. Instead, she ruminates, and then needs to reread her project notes several times because every time she gets to the end of them, she realizes she got lost in thought. She vows again to stay focused.

After ten or fifteen minutes, despite a poor start, she gets into a rhythm and begins making progress. Then an ambulance drives by, with sirens blaring, and snaps her out of her trance. Her mind quickly wanders to the nearby hospital, where she was treated for a broken leg a year earlier, and from there, she recalls the hundreds of times she climbed the steps to her office with a cast on. She is so happy not to have to do that anymore. Her poor mother, that reminds her, is having a tough time walking in general, something Amanda wants to avoid when she grows older. She should really exercise or do yoga, she thinks . . . but when will she find the time to hit the gym or the yoga studio, with all these projects on her plate?

What are you doing, Amanda?! she scolds herself. *What's wrong with you? Focus!*

Ten minutes later, she starts to get back into her project. But only five minutes go by before her partner walks into her office to ask her a "quick question." It isn't quick, of course. The whole time that her partner stands in her office (which turns into twenty minutes), talking about his problem, her stomach acid churns as she feels precious time slipping away.

Through the rest of the day, she struggles to keep her focus. Every time she catches herself being distracted by e-mail notifications (which she immediately opens), by phone calls (which she answers without fail), and by a couple of trips to her favorite gossip website, she reprimands herself and demands that she stay focused on her work until the project is done. By the end of the day, she has only managed to complete one of the three deliverables, and she therefore has to spend more valuable time on the phone with the two clients whose deliverables are not finished, managing their expectations and buying some more time.

Amanda is talented and very good at what she does. But she feels that she has failed to take her business to the next level. If only she could bring in enough business so she could hire someone to do the administrative stuff that robs her of time—like invoicing. She's convinced that she needs more discipline to fight the urges to waste time, and then she'll be able to power through her work without getting sidetracked. Effectively, she demands of herself more willpower in order to stay focused. But she's been demanding that of herself for years, and it hasn't worked.

There's a good reason it hasn't worked, as we'll see. Also, if willpower is not helping Amanda stay focused, what can?

—

ALTHOUGH OUR ABILITY to sustain attention on a task is critical for our success, finding focus—being able to maintain our attention without distraction—is a remarkably difficult thing to do. That's because our brains are actually constructed to respond to distractions. And never before have our workspaces been more distracting: shared offices, meetings, computers, smartphones, tablets, countless e-mails, and the Internet and social media access our devices provide all vie for our attention.

To stay on task, we need to master two skills. The first is obvious—to remove distractions—and learning more about how attention works can help motivate us to take that seriously enough. The second is paradoxical. Of all the strategies for being highly productive I offer in this book, this one is perhaps the most confounding. We have to let our minds wander. That's right. Let that tight grip on attention go.

Before we explore how to do that, let's debunk the theory that to stay on task, all we need is willpower.

THE MYTH OF SUSTAINED ATTENTION

If you struggle to keep your attention focused on a task for a long while, then you are not alone. In fact, your brain is not designed to focus indefinitely on any one thing. It is designed to rapidly switch back and forth between foci of attention. Why? From an evolutionary perspective, it's hard to imagine how we could have survived otherwise. Detecting approaching people, animals, flying objects, and so on is a straightforward survival strategy, whereas staying focused on one thing without keeping an eye on potential dangers would leave us fairly exposed.

Switching back and forth between foci of attention in our environment is also an ingenious way to help us effectively scan our surroundings to find what we are searching for (say, street signs that indicate we are on the right path to our home when we find ourselves walking late at night in a seedy neighborhood) and especially to pick up on what's *changing* around us: something new that we haven't seen or experienced before, or something that violates our expectations (say, when we're driving, a car that unexpectedly cuts across in front of us to make a left turn, even though its blinker suggests it is about to turn right). Parts of our brains are devoted to switching attention—to disengaging and reorienting to a changing environment.[1] For example, if we are reading a menu and we hear the waiter come over to take our order, it is necessary to disengage our focus from the menu and reorient our focus onto the waiter in order to let him know what we'd like to have. This ability to disengage and reorient is adaptive whenever we need to shift our focus.

As your brain switches rapidly back and forth between foci of attention, it gets used to certain inputs that don't change—like, say, that long report you are reading—and starts ignoring them. Your brain first encountered that report twenty minutes ago as it was rapidly switching back and forth between foci of attention around you. It was very happy with its discovery: *Oh, new thing. Let's focus on it!* But as your brain continued to rapidly switch back and forth between foci of attention, the report lost its novelty: it was there in front of you twenty minutes ago and it was there again five seconds ago. Soon your brain became habituated to seeing the report sitting there, a few inches from your face.[2] So it started ignoring it and focusing its attention on other things, primarily anything that was new or different—whether external (a loud noise outside your office) or internal (a memory or future plan that just popped into your head).

In short, our attention systems seem to have been built for scanning and detection, for reacting to the unexpected, for keeping up with what's *changing* around us, and for finding what's new—in other words, for zeroing in on distractions. They do not seem to have been built for being perpetually excited by the same thing, or for blocking out these distractions. Thus, it is wholly unnatural to focus without wavering. If you have failed at maintaining continual focus throughout your work sessions, rejoice. If you had, you'd be remarkably *dys*functional.[3]

In fact, it is so unnatural to focus without wavering that when we try, it backfires. Many people, frustrated by their inability to sustain their attention on a task, try to change their attentional habits by using willpower to force themselves not to give in to distractions. Amanda would have never yelled at a colleague "Hey, you! Just focus!" Nor would a coach shout at a baseball pitcher "Concentrate!" just as he's releasing his pitch. And yet we try this approach on ourselves all the time. We chastise and become frustrated with ourselves whenever we give in to a distraction. We tell (or threaten) ourselves "Don't think about looking at the gossip column," "Don't think about your fantasy football team," "Don't think about the latest gadget you want to buy."

But scientific evidence suggests that doing so is a great way to get *stuck* on a distraction. Studies have found that when people are asked not to think about something, it increases the likelihood that they will think about these things. Don't think about a polar bear right now, and see how that goes. Can you *not* think about a polar bear? Or is it *all* you can think of?[4]

Brains are made up of associative networks of neurons. That is, each neuron is associated with many others. And whenever a neuron is sufficiently stimulated, it will either excite or inhibit the other neurons with which it is connected, and this "activation" will spread across the network of neurons.[5] For example, if you think of the words "polar bear," doing so will activate a network of neurons, probably prompting images of bears, memories of Coca-Cola commercials and childhood zoos, feelings of sadness for the cute and dwindling species, and so on. There is a logical part of your brain that heard the "Don't think" part of "Don't think about a polar bear," but once the network was activated, it was game over.

Since our work demands focus but our brains are wired to be distracted, one of the best tactics to effectively do our finest work is to remove all those distractors we set up for ourselves unnecessarily.

DON'T BOOBY-TRAP YOUR
WORKPLACE WITH DISTRACTORS

Like Amanda, most of us both expect to have the power to focus for hours on end and beat ourselves up when we almost invariably fail to do so. But as we found out earlier, our brains are distraction-finding machines, which makes focusing on a single task for long periods of time quite difficult.[6]

So what can we do to make staying focused easier? Remove the most predictable sources of distraction. Removing distractions in order to focus better may seem incredibly obvious—and it is. Just about every productivity blog post and book recommends doing so, not to mention that it is also just plain common sense. Of course you'll be able to focus better if your colleagues are not stopping by every five minutes to say hi or ask you a question. And yet if you work in an office, then you probably know firsthand that most of us do very little to truly remove distractions from our workspaces. In fact, our work tools—computers, phones, tablets—are incredibly disruptive to the kind of work that most professionals, especially knowledge workers, need to do: think creatively, make complicated decisions, and plan and coordinate tasks.

Our devices help us communicate with others, let us stay in touch with our friends, allow us to share photos with our loved ones, keep us informed and entertained, and make life easier for us in some ways. Of course we love them. But they also play into the brain's natural urge to frequently shift focus from one thing to another. They keep us constantly monitoring incoming e-mails, texts, calls, and status alerts. They offer countless opportunities to get sucked into activities—reading news, playing games, tinkering with apps. They contribute to our mental fatigue by increasing the number of decisions we have to make as we answer e-mails and texts well after the workday is over. And they present numerous opportunities to react emotionally—to somebody's angry e-mail or that sad story on Facebook or that aggravating news about the latest political scandal.

In short, these devices make it really, really hard to be in the mental shape required for good thinking and great work.

Imagine if someone set booby traps around your office—maybe a bucket of water is balanced precariously on top of a door, so it will tip and drench you when you walk in, or a bunch of thumbtacks or a whoopee cushion wait for you on your seat. That's more or less what you're doing to yourself when you set up your devices and workspace so that distractions are coming at you all the time. You have created a work setting booby-trapped not with buckets of water and thumbtacks but with phones, screens, websites, open doors, etc.

If you want to maximize your attention, limiting noises and turning off as many of these devices as possible is a great place to start. There's no need to be a hermit or drop off the grid. Just find a way that your devices can't divert you for perhaps twenty minutes at a time. Close your e-mail application and turn off all notification features in your devices (don't be fooled into believing you'll be able to ignore e-mail alerts when they pop up—your brain won't let you). Close your office door or, if you work in an open space, wear noise-cancelling headphones. Forward your calls to voice mail. Put your devices away or at least a few feet away—make it difficult for you to pick up your smartphone or tablet "just to check if that e-mail came in." If you are working with others in a meeting, lower the screen on your laptop after you finish taking notes so it doesn't distract you from the conversation. Or better yet, rely on pen and paper.

Researchers have found that there are ways of training the brain to sustain attention more effectively too. It may not be for everyone, but if you are so inclined, meditation practice can help. A team at the University of London gave a group of seasoned meditators and a group of non-meditators a sustained attention test. These kinds of tests require people to pay attention for a specific amount of time in order to perform well. In this particular case, study participants heard a series of beeps and had to report how many they heard in each series. Fewer errors would mean better attention. You can probably imagine how boring the task would be, and therefore how much sustained attention would be needed, making it a good diagnostic test. Meditators had better sustained attention than non-meditators, and those with the most meditation practice had the best performance.[7]

Whether or not you enhance your sustained attention through meditation practice, simply removing predictable distractions can increase your chances of having nice blocks of time to get some good work done. But while removing the most predictable sources of distractions from your environment can help you stay focused, you can't prevent *all* distractions. You may be able to turn off the phone ringer, but you can't stop a fire truck from driving past your office window with the sirens blasting.

So what can you do to stay productive when you do get distracted? The answer may surprise you.

LET YOUR MIND WANDER

From a young age, we are taught to believe in the power of sustained attention. It's drilled into us that a good student is one who is attentive all through class, quiet except when asked a question, and quick to follow instructions on a new lesson. Daydreaming is not a skill that we are encouraged to develop. I bet you've never seen a report card with the comment "He doesn't daydream enough."

As adults, when our minds wander, when we catch ourselves drifting off in thought—say, about the game coming up this weekend or what will happen on our favorite reality TV show or whether we remembered to leave a tip at lunch—rather than thinking about the task at hand, we apologize. If our minds wander a lot, we consider it a flaw that we need to manage, something to be embarrassed about. Research, however, suggests that mind wandering may not be a flaw after all. It may have important benefits when it comes to our performing the kinds of tasks that are among the most cognitively challenging to professionals: creative problem solving and long-term planning.[8]

The Benefits of Mind Wandering

Creative problem solving is not something only people in creative fields, like our web designer Amanda, have to do. Regardless of what we do for a living, all of us are faced with issues or problems that we haven't encountered before and for which we need to find unique solutions. From a pediatrician who is trying to decide on the best and safest course of treatment for a tricky case to a manager who needs to design a process for the members of her team (all located in different countries) to effectively communicate with one another, professionals in every industry face complex tasks that require creative solutions.

Most of us assume that the best way to deal with a problem that requires a creative solution is to focus on it relentlessly. But a team of researchers based out of the University of California at Santa Barbara has found that this might not be the case.

In a 2012 study, this research team asked 145 participants to perform what's called an "unusual uses task," which is a test that for decades has been successfully utilized to measure degrees of creative problem solving. It involves presenting participants with a common object, like a bottle, and then giving them a limited amount of time to list as many uses as they can think of for that object.

The participants' performance is then scored based on the uniqueness of the answers.[9]

All the participants in the study began by performing two unusual uses tasks. After they completed those, three groups were given a twelve-minute "break." During the break, participants in one group were asked to do some cognitively demanding work that involved using their working memory. A second group was asked to perform a cognitively easier challenge, known to elicit mind wandering. A third group was asked to rest and do nothing during this twelve-minute break. A fourth group was given no break at all.

Immediately after this twelve-minute period, the first three groups were given a questionnaire, asking them to rate how frequently they focused on thoughts that were not related to the tasks assigned to them (e.g., ruminating about a worry that they had). In this way, the researchers could monitor that mind wandering did, in fact, occur, as they had expected. Then participants in all groups were asked to complete four more unusual uses tasks—two of these were exactly the same as the tasks they had completed before the twelve-minute break, and the two others were completely new.

The researchers found that participants in the second group, who were asked to perform an easy cognitive challenge in between the unusual uses tasks, had, as expected, significantly more mind wandering than those in the first group, who were asked to perform a task with a more demanding working memory load. And sure enough, those in this second group—whose minds wandered the most—were the only participants who did better on the two unusual uses tasks that were the same as the tasks done before the break.

In other words, participants who did more mind wandering got more creative on the repeated unusual uses tasks; they came up with more creative solutions to the problems presented to them after they had some time to let their brains chew on them, so to speak. The other three groups—the one that performed the cognitively demanding work, the one that did nothing, and the one that was given no break at all—showed no improvement on the repeated unusual uses tasks.

It is worth noting that none of the four groups— including the group whose minds wandered most— showed any improvement on the *new* unusual uses tasks. This finding led the researchers to conclude that while mind wandering didn't make the participants more creative in general, it helped them creatively solve the problems they had been working on before they started mind wandering.

The study suggests that if you want to solve a particularly dicey problem, letting your mind wander by engaging in an unrelated and cognitively easy task can help you find some creative solutions to that problem. The UC

Santa Barbara research team even found evidence that people who daydream more frequently in everyday life are generally more creative.

So the next time you find your mind drifting away from a complex challenge or a problem you are trying to creatively solve, rather than yell at yourself for losing your focus—as Amanda did in the opening scenario—just let it happen and reap the benefits of mind wandering.

If that's not enough encouragement for you, consider that mind wandering seems to help with the highly challenging task of long-term planning. I'm sure that sounds paradoxical at first, but mind wandering helps with this because it enables us to think in the right ways about the future.

In a separate study, also led by the UC Santa Barbara lab, study participants were given a task (discerning odd and even numbers as fast as possible when prompted) and a working memory challenge—not to see how well they could perform but to give them a cognitive task demanding enough that they would need to stay focused. At multiple times throughout the experiment, the participants were interrupted and asked to communicate what they were thinking about at that moment. Coders were then able to look at the thought content and identify where the participants' minds had wandered. Most of the time, their minds hadn't wandered to some embarrassing moment in the past. Instead, their minds wandered primarily to the future, and in particular, thinking about themselves and their goals. When their minds wandered, they defaulted to sorting out their personal plans. Had they stayed purely focused, they would have missed out on that important mental work.[10]

When your mind wanders, it's like P. T. Barnum putting on a sideshow while the stage is being rebuilt. Enjoy the show, and when you turn back to the main stage, the next act will be ready to delight you.

What If You Let Your Mind Wander Too Much?

There's a difference between letting your mind wander productively and getting sidetracked. In fact, getting sidetracked is what mind wandering can help you prevent, because it provides a useful alternative to what really sidetracks you.

I see two ways that you can increase your productivity by letting go of your focus in a deliberate way. The first is to actively enable mind wandering: After focusing on a problem for a while, switch to a task that is mildly demanding from a cognitive standpoint—but one that doesn't require you to use your working memory[11]—and then come back to the original problem you are trying to solve.

Choose a task ahead of time, so you don't have to try to remember it once your mind has already started to wander. That will increase the likelihood you'll do it when the time comes. Pick a task that isn't likely to last long (on the order of minutes) or to get you so lost on autopilot that you don't come back. That way you can let your mind wander productively, rather than get absorbed in something else too compelling. The tasks in the following list of examples don't have a heavy cognitive load, don't tend to last long, and probably lose their appeal after a few minutes. Thus, they have a built-in mechanism that will enable your mind to drift away but then also drift back a short time later to what matters. The tasks I recommend include:

- appreciating a piece of art on the wall, a plant in the room, the view out the window, or photographs on your desk, and noticing the different shades of color;
- straightening up your desk, organizing your bookshelf, or cleaning up the whiteboard;
- listening to music and noticing all the different instruments in the piece you are listening to; or
- playing a little game, such as making a mark on a piece of paper every time you see someone walking while texting.

These tasks require a little thinking, but not much. And they don't require much working memory—they don't require you to hold much information in mind while you work through that information.

Examples of tasks I recommend avoiding when you want to do more mind wandering include:

- filing paperwork (often there's a lot to keep in mind to determine where everything belongs);
- reading a sports page, news feed, or blog post (it's too easy to get highly focused on the content, which can block out mind wandering);
- checking and responding to your e-mail (which can both require working memory and grab a lot of attention);
- rehearsing a presentation or preparing for a meeting (memorizing information so you'll have it accessible later can demand a lot from your working memory); and
- working on a tough puzzle, like a crossword or a math game (both can require a lot of working memory).

The second way you can become more productive by letting go of your focus is through engaging in mindful attention. You may have heard of mindfulness-based stress reduction (MBSR),[12] a practice we owe in large part to Jon Kabat-Zinn, who adapted elements of certain Eastern meditation traditions into a structured course for Westerners. MBSR has been shown to be useful for stress reduction,[13] emotion regulation,[14] and fatigue,[15] among many other benefits. I am not suggesting that you take an eight-week MBSR program and then build twenty minutes of meditation practice into your daily routine. While there can be benefits to doing so, I believe there is a lesson captured in the idea of mindful attention that we can start to apply now.

Mindful attention means letting our thoughts go where they want to—that is, letting our minds wander—and, after noting without judgment that our thoughts have drifted, gently bringing our attention back to what we are experiencing in the present moment.[16] Try it with yourself while reading. When your attention drifts at some point, simply note the fact that it drifted as interesting, and gently bring your attention back to the book. It's a way of remaining present most of the time—aware of yourself, others, and your surroundings. When we find ourselves mind wandering, it is possible to become an impartial observer of our wandering thoughts, rather than berate ourselves for getting off task. When we avoid becoming frustrated, frazzled, or further distracted by our inability to focus, we can be more effective at bringing our attention back to the task at hand.

If you've ever surfed or watched surfers in action, you may be familiar with this scenario: After paddling out, surfers will sit on their boards kind of blissed out, bobbing up and down with the waves, waiting patiently—anywhere from seconds to many minutes—for the right wave to ride. Surfers could, in theory, chase after every wave that comes to them. But to have a great session, they let most of them go, until the one that feels and looks right comes along. And the right wave can be truly epic.

Your thoughts are like those waves. When you are

trying to be productive and focus on a task, many—perhaps hundreds—of thoughts will come your way. Mindfully attending to those thoughts means watching the thoughts go by and noting whatever comes up—e.g., whether they stir up worries, tempting you away from the task at hand. The key is to let go of those thoughts that are not helping you stay on track, the way a surfer passes up the opportunity to ride those waves that are not quite right. Brains have lots of thoughts. You don't have to react to each one just because it came up. Be that mental surfer and surf your thought waves.

Quite the contrary to reacting to each thought, if we let those thoughts go, we create the opportunity for our attention to eventually drift back to the original task at hand. In my experience, that's on the order of a few minutes, perhaps as many as fifteen minutes now and again—not a huge amount of time compared to the time lost by starting on a less important project, or by reading the sports news or social media, shopping online, and so on.

If you want to stay on task for a long time, don't fight distractions, but don't blindly give in to them either. When your mind wanders, trust that it needs a minute to do some processing, refreshing, or updating . . . so let it. But don't switch tasks.

If you become distracted by, for example, thoughts of a new diet, consciously notice that thought for a couple of minutes rather than wish it away, but without mindlessly following that thought where *it* wants to take you—to, say, a health website or a blog by a diet guru. After a few minutes of letting your mind wander, I'd wager that you'll be ready to get back on track much faster than you would have had you forced yourself to stop thinking about the distraction—and certainly faster than had you switched tasks and gone to that website.

This, of course, is easier said than done. It is common sense that if you want to stay on task, well, you shouldn't switch to a new one. But we have a lot of practice letting our brains zero in on distractions (as they are designed to do) and then simply falling prey to them. The next time you find yourself daydreaming, tell yourself it's okay, but don't leave the room or turn on any media or start new work like answering an e-mail. Just sit with your drifting thoughts. When your mind wanders, follow; don't lead. You'll probably find yourself coming back to the original work within minutes, and much more ready to really dig in, without needing to fight yourself.

CUT YOURSELF SOME SLACK

Like Amanda, you have probably believed at some point that if you put your mind to it, you ought to be able to stay focused for an indefinite period of time. When you didn't, you may have become frustrated and disappointed in yourself. But as you've seen, you set yourself up for failure when you start with the expectation that you should be able to stay focused without a break. Your brain will do what your brain is designed to do: find distractions and zero in on them. You can limit the distractions and ensure your brain will stay on task for a short period of time, but you can't will your brain to ignore all distractions.

Amanda could have prevented her partner from interrupting her focus by closing the door to her office or letting everyone in her firm know that she would be unavailable for the next couple of hours. She could have turned off the e-mail notification function in her computer to avoid the temptation to open every new message that came in. She could have even installed software that blocks access to her favorite gossip site at certain times of the day. But she could not have stopped as easily her distracting thoughts about her delinquent client or her lack of exercise—or hundreds of other thoughts—from interrupting her focus.

Her only chance to stay on task in the face of these distractions is to let go: to accept these interruptions, allow for some valuable mind wandering, and then gently bring her attention back to her task after a few minutes. And to do that, what she needs more than anything is to cut herself some slack.

It's all too easy to waste time and energy fighting our brains' tendencies to wander, to blame our inability to sustain attention on a flaw in our character. I hope this strategy has shed light on the fact that letting our minds wander is not only normal but also desirable. Our society, though, has so ingrained in us that daydreaming is bad that whenever we find ourselves doing it, we can't help but reproach ourselves for being lazy. That's why, above all, when it comes to staying focused for a prolonged period of time, our secret weapon is not discipline or willpower but self-compassion.

The next time you find yourself daydreaming, be nice to yourself. It may just be what makes the next couple of hours awesomely productive.

—

LEVERAGE YOUR MIND–BODY CONNECTION

So far, we've discussed how to enjoy a couple of hours of real productivity by taking advantage of your decision points, choosing what tasks to take on and in what order so you can bring the right mental energy to them, and managing your attention effectively once you start a task. These three strategies alone could help you make every day hugely productive—if only you had complete control over your calendar.

But, of course, none of us do. We don't always have the flexibility to decide to work on a key project when we have the right mental energy for it. Sometimes we find ourselves not thinking very clearly, being anxious, or generally feeling overwhelmed, but our jobs demand that we perform. Often our daily schedules are filled with presentations that were planned weeks ago, deadlines imposed by our bosses or clients, recurring meetings with our colleagues, and so on. In short, sometimes we want to be at our mental best at specified times and for previously scheduled tasks.

For peak mental functioning in these circumstances, we can add one more strategy: leveraging the *immediate* effects of physical activity and food to improve our mental functioning.

For example, Jennifer—who through the years worked her way up to being the head of human resources for a ten-thousand-plus-person organization—has to meet the Japanese owners of the parent company, who are in town. Twenty minutes before the meeting, her stomach is so knotted up from drinking too much coffee and eating the candies she keeps at her desk, and her neck muscles are so sore from stressing out over reviewing the agenda on her computer again and again, that she is highly distracted and irritable.

Ten minutes before the meeting, she finds herself standing in the bathroom, wishing this hour could just be over. Staring in the mirror, she notices the bags under her eyes that weren't there that morning. *Do I really look that old?* she thinks. Her body is broadcasting on the outside what she feels like on the inside. However, as you'll learn in this strategy, had she gone for a brisk thirty- to forty-minute walk instead of obsessing over her meeting materials, her body could have had a lightness of spirit and mental clarity to broadcast instead.

That our physical states influence our mental states is at once obvious and revolutionary. On one hand, it is obvious because almost everyone has experienced, at some point, how our minds feel sharper when we feel physically great or pretty foggy when we are sick (or when we're suffering from a food coma). These are the effects of what's happening in our bodies on our capacity to think.

On the other hand, it is revolutionary because we seldom act on this intuitive knowledge. That's likely because the idea of the body and the mind influencing each other flies in the face of age-old assumptions in Western thought. For a long time, the mind was treated as though it were independent of the body—and the body was treated perhaps as only a life support system for the mind.[1] No wonder we expected people to function equally well all day long and acted like exercise was a luxury.

Recent research, however, confirms and helps us act on what we know intuitively—that our physical states affect our mental states deeply. The mind and the body are so intertwined, in fact, that we often confuse how we feel physically with how we feel emotionally—and this can be an incredible tool when setting up two awesome hours of productivity.

In a well-documented study, done in the 1960s, two researchers—Stanley Schachter and Jerome Singer—gave study participants an injection of adrenaline, telling them it was a vitamin solution.[2] Some participants were told to expect side effects from the shot, such as increased heart rate, trembling hands, and a flushed face—in other words, the actual side effects of adrenaline. But other participants were either not told there would be any physical side effects or given a list of fake side effects they could expect.

The participants were then exposed either to a situation that was likely to elicit elation or a situation likely to cause anger. Those who had been told to expect the actual side effects from the injection more often identified those side effects as physical sensations. But those who had not been told about the actual side effects tended to experience the physical symptoms—the flushed face, the shaking, and the increased heart rate—as emotions (elation or anger, depending on the situation they were exposed to).

What Schachter and Singer showed was that it can be very difficult to distinguish a physical feeling from an emotional one. The two are thought to be tightly connected. Your emotions can feel just as real to you, regardless of whether they were elicited by a situation or by a physical response to a substance, such as adrenaline. Since that is the case, then changing what your body feels like can be a way to help change your mental state.

Few of us find ourselves in a situation in which we've just received an injection of adrenaline. However, is it possible that caffeine could from time to time induce you to feel different emotionally? Or might you mistake the physical effects from eating a heavy carb-packed meal with irritation toward a coworker?

In this strategy, I won't make an argument that you

should be in top physical shape in order to be more productive (we'll leave that for another book). If you like, go ahead and eat a gigantic carb-heavy lunch, slump at your desk for hours on end, and avoid regular exercise. But save that behavior for the hours when you are not planning to tackle tasks for which you need to be in top mental shape.

Instead, for this strategy I'll explain how physical exercise, food, and caffeine can influence your mental functioning in the minutes and hours that follow. When you find yourself struggling to think clearly, feeling overly anxious, or feeling overwhelmed, it's time to use your body differently as a way to hit the reset button. By understanding how exercise and food affect your mental functioning, you can use them as tools to help you be more productive when work demands it—for example, when you are about to give a presentation, working to meet a deadline, or getting ready for an intense meeting with a client. Of course, any time you change your eating or exercise patterns, you should check with a doctor first.

HOW EXERCISE AFFECTS YOUR MENTAL STATE

There is no doubt that many personal attributes made Nelson Mandela capable of immense mental resilience during his years in hiding and then his decades in jail as a political prisoner. But he attributed his clarity of thought and resilience in part to his physical activity—even when he was confined to a cell day after day. In his autobiography, Mandela revealed that while he was in jail, from Monday through Thursday he used to run in place for a maximum of forty-five minutes as well as do push-ups, sit-ups, and other exercises. "I found that I worked better and thought more clearly when I was in good physical condition, and so training became one of the inflexible disciplines of my life," Mandela wrote.[3]

For decades health advocates have been urging all of us to exercise more. The long-term benefits of consistent exercise to our health, well-being, and looks are well documented, and surely you've heard them several million times, so I won't be repeating them here. In this strategy, I will highlight the message Mandela seemed to have understood very deeply: that physical exercise has a near-term effect on mental performance.

There are immediate benefits of exercising—which can occur after a single session of activity—that you may not hear about from health and fitness sources and that pertain to your mental state. Even a little exercise at the right time can help you think better, stay focused, sharpen your thoughts, and reduce your anxiety—key elements of sustained productivity—in the *hours* that follow the physical activity.

For example, one meta-analysis showed that exercising for ten to forty minutes has a consistent and immediate effect of improving executive functions.[4] You may remember from strategy 2 that the term "executive functions" refers to the brain's various abilities to direct and override other mental activity, like prioritizing some items over others at a staff meeting or stopping your impulse to call the boss stupid in the middle of your performance review. Research suggests that physical exercise particularly enhances the executive functions that have to do with self-control.

In another example, a group of researchers in Japan asked participants in their study to take a common psychological test called the Stroop test. The particular Stroop test they assigned participants is known as a color-word test: participants are presented with the written word for a color (for instance, "yellow") but the word is printed in a color that is different from the written word (for instance, black). Participants will say either the word or the color in which that word appears. (For an example, as fast as you can, say out loud the color of the ink of the following word: YELLOW. Chances are you'd be slower to say the correct answer, which is black, than if this were the word BLACK.) In the color-word test, a shorter reaction time for correct answers is considered a measure of better inhibitory control.

After taking the test, some participants were asked to exercise for ten minutes at a very precise moderate level of exertion (at half the participants' maximum oxygen intake—a measure that reflects the aerobic physical fitness of an individual[5]—which is roughly like taking a brisk walk or going for a light jog). Then those participants

were asked to rest for fifteen minutes before taking another Stroop test. Meanwhile, a control group was asked
simply to rest for twenty-five minutes (they did no exercise but waited the overall same amount of time) before
taking another Stroop test.[6]

The researchers found that not only did the ten-
minute session of moderate exercise lead to significantly
faster reaction times on the Stroop test, but also part of
the lateral prefrontal cortex—a region of the brain that
plays a role in self-control—showed more activity. These
brain findings suggest that exercise doesn't just make us
more alert or ready to do everything faster; it may also
enhance brain processes responsible for stepping in and
exerting self-control, thereby benefitting the decisions,
plans, solutions to problems, and so on that make use
of self-control. These are precisely the kinds of cognitive
functions talked about in strategy 2 that suffer when we
get mentally fatigued.

There's also evidence that exercise results in sharper
attention. A research group at the University of Illinois
at Urbana–Champaign asked participants to exercise at
a moderate level on a treadmill for twenty minutes (at
60 percent maximum heart rate, which would again be
something like the intensity level of a brisk walk or light
jog for many people).[7] After the participants' heart rates
returned to normal—usually around twenty or twenty-
five minutes after they had finished exercising—the
participants engaged in something called a flanker task.

In a flanker task, participants have to ignore distracting stimuli on a screen that flanks either side of a target they are looking for.

The Urbana–Champaign team found that moderate exercise not only helps to sharpen attention but it may do so by fine-tuning very early attention to incoming information. This suggests that a little exercise can help you focus or concentrate, while also avoiding distractors, at a basic perceptual level. In the work world, that may be similar to, for example, focusing better on the words in a document you are reading on the computer rather than getting sidetracked by all the pop-up windows, alerts, and so on from your devices.

But perhaps one of the greatest benefits of exercise to our productivity is that it helps our overall mental state immediately after we engage in it. Have you ever gone too long without eating and had strange things happen to your mood, your clarity of thought, your ability to focus, or your mental quickness? You were likely experiencing these changes in your mental state because your blood sugar had dropped.[8] We all react differently when our blood sugar drops, but it's not uncommon to experience these productivity-sucking symptoms.

It turns out that exercise helps to stabilize blood sugar levels too. Among type 2 diabetics—people who are at risk for dangerously high blood sugar levels—a single aerobic exercise session was found to drop blood sugar by 16 percent and keep it lower for about three hours.[9] Exercise is a common aspect of treatment for diabetics, and its benefits for them have been known for some time. What that study helps to show is the specific and immediate impact just one exercise session can have on blood sugar levels—and on mental abilities.

Exercise is also fantastic to reduce anxiety. A meta-analysis—analyzing results from over one hundred studies—showed that aerobic exercise in the twenty-one to thirty-minute range was enough to reliably reduce anxious feelings in the period after exercising.[10] And another study showed that in the longer term, exercise actually buffers against the negative effects of chronic stress.[11]

A different meta-analysis also found evidence that exercising has the effect of amplifying positive moods and emotions (like joy, elation, vigor, or enthusiasm),[12] not just alleviating the negative emotion of anxiety. Positive emotions reliably increase after low or moderate exercise, but, surprisingly, not as much after more intense or longer sessions of exercise. The researchers roughly categorized "moderate exercise" as twenty to thirty minutes of a high-intensity workout or thirty to forty minutes of a moderate-intensity workout. Think of high intensity as jogging at a decent pace, breathing heavily, getting your heart rate up a bit, working up a good sweat, or more. Moderate intensity for many people, as I stated previously, is the equivalent of a very brisk walk or light jog, in which you work up a bit of a sweat, loosen up the aches and pains, and need to breathe a little more heavily than usual, but you certainly do not push your limits.

The research, in addition, suggests that this positive effect on emotions peaks within thirty minutes of exercising. What's more, the benefits on emotions and mood were strongest for those who had relatively lower positive moods before exercising. So it seems that exercise is most useful when we need it most.[13]

Still another meta-analysis found that moderate aerobic exercise, lasting somewhere between twenty-one and forty minutes, made people feel more energetic afterward as well.[14]

So what does all this research mean for setting up two awesome hours of productivity?

Exercise Strategically

As I wrote earlier, this strategy is not an argument for why you should exercise regularly for better overall health and perhaps, through that, become more productive. Rather, I'd like to suggest that, whether or not you currently have an exercise routine, you can use physical activity at specific times in order to boost your thinking abilities and your mental energy.

Do you have to give a presentation, complete an important project, craft a strategically important document, or deliver a key customer proposal? Do you get nervous when interacting with clients? Do your check-in meetings with your boss or certain clients raise your anxiety level? Does engaging in certain tasks—like doing tedious jobs, engaging in activities you feel you are not good at, or working with people whom you find difficult—bring down your mood? Do you have some multi-hour meetings after which you feel completely drained? Are there times of the day or week when you routinely feel tired or disengaged? Some moderate exercise at the right time can make the difference.

As you've seen, quite a bit of research suggests that moderate exercise can help you focus, clarify your thinking, and improve your mood (while helping you chill out, too) in the minutes and hours that follow. Exercise is like a reset button. It is a reliable, effective, and fast-acting strategy for improving your mental performance. Moderate exercise—vigorous enough for you to work up a sweat but not to feel spent—is a game changer in the hours immediately after you exercise.

Here are some ways to leverage exercise well:

- If you are feeling mentally sluggish and unable to focus, get out of your office and move right away. Walk very briskly for thirty to forty minutes. Go up and down the back stairs for ten or twenty minutes. Or if you belong to a gym nearby, take a break and exercise for twenty to thirty minutes on the tread-mill, exercise bike, or your preferred machine. Try to break a sweat, but don't overdo it. The moderate

physical activity may just sharpen your focus and
mental agility.

- Whenever possible, schedule challenging or anxiety-
provoking meetings when you can block out time
beforehand for moderate exercise. The exercise is
likely to calm your nerves and improve your mood.
- When there is a particularly challenging or draining
task on your calendar, either exercise in the morning
before it (to make it easier to handle) or schedule it
at a time of day when you can exercise soon after it,
to restore your drained mental energy and improve
your mood in time to tackle whatever comes next.
- In general, plan to work out for about twenty to
forty minutes within a couple of hours *before* you
next need to be awesomely productive.

If Nelson Mandela could run in place in his jail cell,
you can get to a treadmill in the morning before your big
meetings. So the next time you need to tackle an important task, consider whether your body, and not just your
brain, may be the key to your success.

HOW FOOD AND DRINK AFFECT
YOUR MENTAL STATE

If your schedule doesn't allow you to fit in physical activity on a particular day when you most need it, there are

other ways to work in concert with your body to achieve peak productivity. One of them involves something you already do every day, even at work: eating and drinking. What you eat and drink—and when—can have meaningful effects on your energy level, mood, and executive functions in the minutes and hours that follow. Besides recharging your body with exercise, you can fight mental fatigue when you need to be productive by being strategic—changing the content and amount—about what you eat and drink.

The Scoop on Eating Carbohydrates, Proteins, and Fats

There's not a huge amount of research comparing the effects of carbs, proteins, and fats on how we think and feel in the hours after eating.[15] But what there is so far may surprise you. A recent review suggests that carbs can give a very brief, minutes-only bump in some mental capacities (attention, for example, improved fifteen minutes after eating carbs in one study, but by an hour after eating carbs, there were drops in other executive functions). Some believe that given the time it takes to absorb nutrients, effects occurring just minutes after eating probably have to do with something else besides nutrients being absorbed, like just detecting that nutrients are coming. By contrast, a high-protein meal boosts memory an hour afterward.[16]

Meanwhile fats, believe it or not, may turn out to be more helpful. One study found a greater benefit for several executive-function tasks when people consumed fats instead of proteins or carbs (each was in the form of a vanilla cream that tasted similar regardless of macronutrient content), lasting for three hours. The fats in that study (a combination of soybean oil, palm oil, and double cream) did not cause as great a change in blood sugar levels or in the balance of hormones that regulate those levels as carbs and proteins did. Carbs and proteins also resulted in different effects: carbs gave a bump in short-term memory compared to protein, and protein gave a boost in attention compared to carbs.[17]

So if what you're after is cognitive performance, maybe don't worry so much about whether you put a lot of cream in your coffee or that piece of cheese is high in fat. (But note that, in the long term, saturated fats appear to be bad for cognitive function, while omega-3 fatty acids seem to be a plus.[18])

Blood Sugar Levels and Productivity

When it comes to eating carbs, there's more research about their effects (as compared to proteins and fats) on mental capacity. Within carbs, there is an important distinction to make. Their effects are stronger or weaker depending on their glycemic index, a measure of how much a carbohydrate raises our blood sugar levels. When sugar circulates in the blood, it is in the form of glucose, a major fuel of the brain and body. If we eat straight glucose, it increases our blood sugar sharply. Glucose is assigned a glycemic index of 100 to mark the top of the scale. Other carbs have a weaker effect on blood sugar. For example, an apple has a very low glycemic index of just 34—meaning its effect on our blood sugar levels would be 34 percent of the effect a comparable serving of pure glucose would have.[19]

Most whole vegetables and fruits, in fact, are lower on the glycemic index than breads, cereals, pastas, cookies, cakes, sweets, and sugar. There are a couple of exceptions: Oats and quinoa are lower on the glycemic index than many other grains, and bananas are higher than many other fruits. And this one sometimes trips people up: juices may be made from fruits low on the glycemic index, but multiple pieces of fruit need to be pressed to make juice. As compared to eating the whole fruit, a little bit of juice is often a lot of carbs.

Maintaining a stable blood sugar level is best for cognitive performance and a stable mood. So the kinds of carbs we eat make a difference too. In one study, with three groups of participants, each group ate a breakfast including bread with a low-calorie jam, low-calorie yogurt, and an orange-flavored drink, but the glycemic index of the yogurt and drink servings varied: 100, 67, or 32. The participants' moods were then assessed for a few hours after their meals.

The researchers found that the higher the glycemic index of the breakfast, the more hostile or disagreeable the participants were. However, as the authors of the study pointed out, the story is more complex: mood and cognitive effects also depended on things like a person's ability to metabolize glucose well. Surprisingly, participants whose bodies metabolize glucose well saw more effects, while those who don't manage glucose well didn't display as many differences.[20]

Another study with twelve- to fourteen-year-olds found that a low glycemic index breakfast helped with executive functions, for example on a Stroop test (self-inhibition) and a flanker task (attention in the face of distraction), as compared to either a high glycemic index breakfast or no breakfast, and the primary benefits occurred two hours after the meal. In that study, the high glycemic breakfast was in the low 70s on the glycemic index and the low glycemic breakfast was in the high 40s.[21]

How Much Should You Eat at Once?

It's not just what we eat that can affect our energy and mental sharpness. How much we eat *at once* matters a great deal too, even within a normal quantity range. In a study conducted in the UK, researchers served two groups of people precisely the same food but gave one group the food in two larger servings and the other in four smaller servings. The participants having just two servings drank milk shakes with a macronutrient content of 58.4 grams of carbohydrate, 21.5 grams of protein, and 25.2 grams of fat at nine A.M. and then at one P.M.—like the timing and size of a reasonable breakfast and lunch. The other group of participants drank four milk shakes

with half the macronutrient content (29.2 grams of carbohydrate, 10.9 grams of protein, and 12.6 grams of fat) at
nine A.M., eleven A.M., one P.M., and three P.M.

An hour after consuming the milk shakes, both groups
engaged in tasks designed to test reaction times, verbal
reasoning, memory, and more. Participants who drank
the four smaller-size milk shakes made significant improvements in a number of these tasks. The researchers
speculated that smaller meals spaced out help regulate
blood sugar levels, which in turn has a positive effect on
thinking, in particular on working memory.[22]

Drink Up

What we eat has important implications for our mental
energy—and what we drink does as well. Let's start by
taking a look at the most popular drink—water—and
what happens when we don't drink enough of it.

Water

Roughly 50 percent of the human body is estimated to
be water.[23] We depend on water in numerous ways for the
basic functions of life. It only makes sense that keeping
us, and our brains, hydrated is critical for our performance and our ability to set up two awesome hours of
productivity. But dehydration is a problem that probably
affects us more often than we think. Even mild dehydration can have a negative effect on our mental energy and
our ability to be at our best.

A review of the literature looking across a number of studies shows that even among healthy young adults, just a 2 percent drop in hydration can hurt attention and short-term memory, although not long-term memory or a number of executive functions.[24]

And in terms of subjective experience, dehydration seems to make things tough too. For example, one group of researchers explored the effects of dehydration by having women walk on a treadmill for forty minutes. On average, the women lost 1.36 percent of their body mass due to dehydration. On some test days the women replenished their water loss by drinking, and on other days the same women did not. When they were dehydrated the women were angrier, more fatigued, and felt it was harder to concentrate.[25]

Moreover, as we become elderly, hydration seems to be more relevant to cognitive performance and better mood.[26]

Don't overlook the power of a simple glass of water to keep your mind sharp. If you need to be on top of your mental game—and you haven't drunk water in the previous hour or two—head to the water cooler right away. Or why not add a little reward to the experience by replacing a plain glass of water with something fancier? I'm a huge fan of seltzer—or, as my nephew dubbed it when he was three, "spicy water."

Caffeine

There is another drink to which many of us turn to fight fatigue: coffee. Some estimate 80 percent of people around the globe drink coffee or tea daily, or consume some other source of caffeine.[27] If you are like me, you have come to depend on coffee in the morning to help get your brain going. But also, you've probably noticed that there are times when you drink it and it just doesn't seem to work. It's like the coffee's broken.

Caffeine—in the form of coffee, tea, sodas, energy drinks, or hot chocolate—is a mixed blessing. It sometimes helps and sometimes hurts our cognitive performance and mood. Recent research suggests it is a net plus for many people, when used properly. Let's turn to some of that science to understand what proper use means and why caffeine has an effect on us.

Some researchers believe that the benefits of caffeine to our mental faculties actually come once we are dependent on caffeine. When we are not drinking caffeine—for example, during the long stretch of time when we are sleeping at night—we go into withdrawal. During withdrawal, our mental faculties become diminished, we may be in a relatively negative mood, and we may even get headaches. Once we have that cup of coffee or strong tea, the surge of caffeine in our system makes us feel wonderful in comparison to how we felt while in caffeine withdrawal. It sharpens our minds and improves our moods by comparison. As the caffeine flows through our veins, it brings us back to baseline, to a level of functioning that we might have experienced, all else being equal, had we not become dependent on caffeine.[28]

But don't worry. That doesn't have to mean you should wean yourself off coffee. Coffee has been shown to have positive long-term benefits on mental and physical health, such as slowing cognitive decline as you age and decreasing the risk of developing type 2 diabetes.[29]

Other researchers believe that caffeine affects our mental faculties whether we are dependent on caffeine or not. For example, a study out of the University of Chicago looked at the effects of caffeine on participants who drank less than 300 milligrams (about three 8-ounce cups of coffee—or one Starbucks Grande) a week and had no coffee dependence. Thus, the effects could not be due to overcoming withdrawal. The study found that 150 or 450 milligrams of caffeine increased stimulation, decreased fatigue, and led to better attention, as compared to placebo. However, it wasn't all positive—450 milligrams also increased anxiety and hurt memory somewhat.[30]

Both theories are probably right: some of the apparent benefits of caffeine are probably a result of reversing withdrawal symptoms, and some are probably a direct effect of the drug itself. Either way, research suggests that a little caffeine will probably put you in a better mental state to be productive. But as with any drug, it matters how you take your dose and how large your dose is.

If you are looking for a mental boost, it may be best to have that dose of caffeine with some food. In one study, researchers found that drinking caffeine accompanied by just water (as in black coffee) can give you a boost in how you feel thirty minutes after ingesting it; but an hour and a half to two and a half hours later, the caffeine can make it harder for you to feel like you can think clearly, make you feel more tired, and lead you to feeling more hostile.[31] However, mixing caffeine into a yogurt drink instead of into water reversed these effects, making the positive benefits in terms of how people felt last throughout that time. Since many people take coffee with sugar alone, it may be helpful to note that sugar was not enough; just mixing the caffeine with glucose, in that study, did not help with how people felt in the hours after having it. The body is a master chemist, and it can combine foods or drugs in ways that end up making a meaningful difference in our experience.

So how much caffeine is necessary to reap benefits in your productivity? It depends on the person. You have to figure out the dose that is right for you—and then stick to it. As the physician Paracelsus said in the early 1500s, "All things are poison and nothing is without poison; only the dose makes it that a thing is not a poison."[32] Taking more than the right dose for you can result in negative consequences.

Generally, at low doses, caffeine can lead to a more positive mood and even decreased anxiousness.[33] But some people differ genetically in whether caffeine makes them more or less anxious[34]—you probably know who you are by this point in life from trial and error. If caffeine does make you more anxious, or if it upsets your stomach, experiment with much smaller doses than you've had in the past. The bottom line: research suggests that when it comes to caffeine, less is sometimes more.

One study, exploring attention, found that up to 200 milligrams of caffeine—like a strong medium-size cup of coffee—improved some aspects of attention, but 400 milligrams of caffeine had no additional benefit.[35] Another study compared people who averaged about 150 milligrams of caffeine a day to those who averaged about twice that. Both groups consumed a strong dose—400 milligrams, like the amount of caffeine in a Starbucks Venti (20-ounce) coffee—and both groups had some level of increased anxiousness and tense negative mood afterward. Only the group who typically drank a lot of caffeine reported a benefit in terms of wakefulness with 400 milligrams caffeine.[36]

In a third study, researchers gave study participants 100 milligrams of caffeine (like a small cup of coffee) in the late morning and 150 milligrams caffeine in the early afternoon, mimicking a common caffeine routine for many people. Among their participants, people who did not consume caffeine on a regular basis—or only consumed a small amount (less than 40 milligrams a day)—did not feel more alert or experience enhanced cognitive capabilities. By contrast, those who regularly consumed more than 40 milligrams of caffeine a day did experience an increase in alertness and cognitive capabilities. The authors of the study argued that the anxiety the caffeine caused in those people who don't regularly drink a lot of coffee counteracted any mood or cognitive benefits.[37]

Finally, there is evidence that half a can of Red Bull energy drink (about 40 milligrams of caffeine, along with a number of other ingredients that may influence its effects) has a more desirable effect than either a whole can or a can and a half. The *smallest* dose of energy drink was better in terms of fighting fatigue and enhancing at least one executive function: self-inhibition.[38]

Caffeine should take thirty minutes to reach its full effect.[39] So make sure you wait until it has finally kicked in before you drink some more (a temptation that is common, especially when you are tired or stressed out). Otherwise, you may accidentally overdo it, and instead of enjoying the benefits of a nice mental boost, you may get a bad case of the jitters.

To get the best results from caffeine, and if you're a regular, dependent user, stick to your normal dose even if you feel tired or stressed. Give it time to kick in before consuming much more than usual. And if you're up for trying something new or you're not a dependent user, don't drink it all the time. Instead, reserve it for those times when you are tired or having trouble paying attention due to fatigue. If that doesn't give a boost in attention, alertness, mental energy, and positive mood, drinking more is probably not the answer. You may be better off with a nap rather than more coffee in those circumstances.

Eat and Drink Strategically

We all intuitively know that eating too much or drinking sugary drinks can actually make us sluggish and tired hours later. At some point we've all felt the overwhelming need for sleep that follows an overly indulgent lunch.

So if we know this, why do we do it as often as we do? Many of us are not aware how intertwined our bodies and minds are. When we're not thinking, we tend to go with our habits—and we often have the habit of ignoring the immediate effects food and drink will have on our cognitive skills.

When you want to be firing on all cylinders, being intentional about what you eat and drink—and how you want to feel an hour later—can make all the difference. Here are some key tips for how and what to eat and drink to increase productivity over the two to three hours after your meal or snack:

- Eat only half your breakfast or lunch and enjoy the second half a couple of hours later.
- If you need a quick boost, a high-carb snack may help you focus and feel good for about fifteen minutes. If you need to be in top mental shape for longer than that, avoid carb-rich meals and snacks altogether: Don't go for pasta dishes, sandwiches, or pizza. Leave out the juices, sodas, and sweetened iced teas. Skip the fries, chips, extra bread, and sweets.

- Eat meals or snacks that have a nice mix of proteins, low glycemic index carbs, and good fats—vegetables and fruits are generally good carbs; nuts make great snacks for when you are on the run.
- Don't be fooled into eating a carb-rich meal. The whole meal or whole snack should be high in protein and low glycemic index carbs. If you have some chicken along with a giant plate of rice and beans and a sweetened iced tea, you are eating a carb-rich meal.
- Drink water if you haven't had any for the last hour or two or if you've done any physical activity. It makes a difference.
- If you are tired or sleep deprived, drink a caffeinated drink, but keep it small. Don't drink more than what you normally would. Give it thirty minutes to kick in. And go ahead and put cream in that coffee—the fat may help keep your blood sugar level more stable.

TAKE CONTROL OF HOW YOU FEEL

Typically, when making decisions about what we'll eat, most of us only consider two things: Will it taste good? (That is, will it bring us pleasure?) And is it healthy? (That is, is it compatible with whatever weight loss or health diet we are on that week or month?) Similarly, when we engage in exercise, we typically do so in order to increase our overall well-being and health. But we seldom make decisions about what we'll eat or when we'll exercise based on how we want our brains to operate in the *immediate* hours after the meal or physical activity.

It's easy to think that you have to put up with and learn how to be productive regardless of the way you feel physically—whether you are feeling groggy and unfocused or energetic and clearheaded. But the truth is that you have more control over how you feel physically than you may realize.

Whether you are already in great shape physically or rarely exercise, and whether you are devoted to healthy nutrition or more of a burger-and-fries kind of person, I hope this strategy inspired you to think of exercise and food in a different way: as a tool to help you set up a session of superb productivity.

—

MAKE YOUR WORKSPACE WORK FOR YOU

*S*ix months into her new position, Samantha feels like she is swimming upstream and getting nowhere. She was recently appointed CFO of a start-up and is feeling overwhelmed by the long list of tasks her new boss, the CEO, has assigned to her. Every week brings fresh challenges, from dealing with new acquisitions to overseeing the merging of separate departments' disparate accounting practices to hiring new people to finding new ways of cutting down expenses across the company.

Today, Samantha is determined to make progress on her tasks. So after a meeting with the CEO, she resolves to hunker down at her desk and get some serious work done. Easier said than done. Like many start-ups, her company has an open floor plan. All managers work on the floor with their teams, at moveable desks. As she heads back to the floor, she tries to keep her head down and avoid eye contact; already her needy coworkers are hoping to get her attention.

After being stopped by a couple of team members who have questions or need her approval, she reaches her desk. It has been a rough couple of weeks and clutter has piled up in her small area. Sitting under the dim light of her desk lamp, staring two feet ahead at the optional beige cubicle divider she requested, she nudges a stack of papers aside to find space for her coffee. Needing to think creatively and desperately wanting to focus, she tries to shut out the chatter and noise around her—colleagues talking, phones ringing, printer churning out countless copies of documents. She leans closer to her computer, so she at least won't see anyone coming her way, and sets her elbows on her desk to support the weight of her heavy head in her hands. *I can't get any work done here,* she thinks.

Many of us have seemingly little control over the space in which we work. Unless you are self-employed or work from home, when it comes to your office environment, you are probably at the mercy of your employer's office design decisions. Yet, as challenged as you may be in either selecting or setting up your workspace, I'll share a number of things that are likely to be within your control—whether you work in an office or from home— that can help you set up conditions for performing at your best. (In case you are concerned that my advice won't be feasible for most professionals, don't worry. I won't be suggesting you remodel your office or find a way to work exclusively from home—I know you likely can't.)

Like exercise and food, our surroundings affect our brains in meaningful ways. Learning to manipulate these surroundings so that we can operate at peak productivity levels involves understanding how and why our brains and minds react to external stimuli, and therefore how to successfully work around those stimuli.

I will share some of the most relevant science on how we react to typical workplace stimuli—especially to noise, light, and the immediate physical space in which we work. With this information, we can become aware of how our surroundings may be sabotaging our productivity and what is within our control to address, in order to be in top mental shape when we need it most. The final strategy for setting up two awesome hours of productivity is to make informed choices about our work environment, choices that enable us to focus and do our best thinking.

PRODUCTIVITY AND CREATIVITY IN A NOISY WORLD

Is it better or worse to listen to music while working? What about white noise? Sometimes we don't want to work but have to, so we may decide to work in front of a TV to make ourselves at least feel better about it. Are we just fooling ourselves into thinking we can work in noisy environments? I have bad news and good news. The bad news is that researchers have found that environmental noise—background music, city sounds, people's conversations—leads to a decrease in performance for most people. The good news is that this represents an easy way to make changes that can set us up for being highly effective.

Of the different sources of noise, there is one that turns out to be the hardest to tune out. Intermittent speech is particularly challenging to ignore. Intermittent speech is when you hear a few words or sentences here and there with pauses in between. Like when colleagues who sit in the cubicles behind you turn and ask each other questions, or when someone else is on a phone call listening for a while and then speaking periodically. Intermittent speech is also one of the most common sounds in an office. One meta-analysis examined 242 studies on how noise affects performance and found that when it came to performing cognitive tasks—like staying attentive, reading and processing text, and working with numbers—performance was more affected by intermittent speech than by either continuous speech (which would have little variation in volume and rhythm) or nonspeech noise.[1]

Intermittent speech in the background may be the biggest problem for work performance, but that does not mean other noise (e.g., continuous speech, music, or white noise) is fine. A second meta-analysis looked specifically at the effects of listening to music in the background on performance. While researchers found that listening to background music tended to improve positive emotions, increase performance in sports, and make people do what they do a little faster, it also had disruptive consequences on reading.[2]

If you can't avoid a noisy environment, should you play white noise in order to drown out the rest? White noise is a nondescript background hum, kind of like the noise of a fan or someone saying "Shhhh!" continuously. Listening to white noise may turn out to be better than listening to intermittent speech if it successfully drowns out the speech, but that doesn't mean it's ideal. Quiet should be even more effective than white noise. For example, in one study, the majority of kids (excepting those with the most severe attention problems in the classroom according to their teachers) in a middle school setting had worse memory in the presence of white noise as compared to no noise. Those with the most severe attention problems actually did better with the white noise.[3]

So are all those people working or doing homework in busy, loud coffee shops fooling themselves into thinking they are being productive? I love the ambience of coffee shops, so it is sad for me to say that for the most part, yes, we are fooling ourselves, but with a couple of exceptions.

A lab in Glasgow conducted research to find out if noise affects cognitive performance in introverts and extroverts differently. And it turns out it does. These researchers exposed participants to different types of noise (everyday noise, music that they categorized as having "high arousal potential and negative affect," and music they categorized as having "low arousal potential and positive affect") as well as to silence while the participants completed a battery of cognitive tests.

All participants did worse on the cognitive tests when there was any kind of noise in the background compared to when they performed the cognitive tasks in silence. But the researchers found that participants who were introverts had even more performance problems than extroverts. They theorized that introverts, who are generally more easily overwhelmed by stimuli, are more sensitive to the noise distractions.[4]

But extroverts are not the only ones who have an advantage when it comes to fighting off the distractions caused by background noise. There is also evidence that people who naturally have better working memory capacity—people who are more capable, for example, of remembering a phone number until they dial it or keeping track of what a conversation is about while they are talking—can also withstand background noise with fewer consequences.[5]

So if you are an extrovert or have a good working memory, you may be better able to handle the distractions in a noisy environment. You may find it easier to get that document drafted while listening to music through your headphones or to complete your presentation while listening to your colleagues talk on the phone with clients or to finish that financial report to the whirring and humming of the copier machine that sits right outside your workspace. But make no mistake: you would be more productive if you were doing your work in a quiet environment.

Still, noise is not always bad, and it may provide some benefits to certain kinds of work challenges under the right circumstances. In a study from the University of Illinois at Urbana–Champaign, participants were asked to work on a creative challenge while listening to one of several levels of noise loudness. In the creative challenge, participants were told to come up with as many unique uses for a brick as they could imagine (doorstop, hammer, table centerpiece, and so on). When participants worked on the creative challenge while listening to low noise (at fifty decibels—about the noise of a typical large office), they tended to be less creative than when they worked on the challenge while listening to moderate noise (at seventy decibels—a little quieter than the sound of a vacuum cleaner that is ten feet from you[6]).

It turns out that as the noise level increased, participants had more difficulty thinking. And the more difficulty they had thinking, the more abstract and "bigger picture"—in short, creative—their ideas were. Interestingly, higher noise levels (at eighty-five decibels—like a diesel truck driving by[7]) made thinking so difficult that the benefits of some noise on creativity went away.[8] So this is evidence that moderate noise may help creativity. And, by contrast, too much or too little noise may hurt it.

The research on the consequences of noise on productivity is fairly straightforward: for the bulk of the tasks performed in the knowledge economy, quiet is almost always better than noise.

These are some basic things you can do to help you stay focused when it's important to be on:

- If your office has a door, close it. If you don't have a private office, reserve a conference room or set up camp somewhere that is largely free of noise and other potential distractions. A place with privacy that is away from noise distractions will be more favorable to productivity.
- If your desk is in a shared space and you must stay there, put noise-cancelling headphones on. Those little squishy orange earplugs can sometimes do the trick too, and you can take them anywhere. You may look weird, but you'll be more productive.
- Don't listen to music or talk radio.
- If you are working at home, turn off the TV.
- If you're taking on a task that requires lots of creativity, enjoy background noise. You may actually consider heading for the company's busy cafeteria or a local coffee shop, or putting on a little music.

To get away from noise, there's another alternative that can make a big difference. In so many instances, people will get up early, stay up late, or make time on the weekends to get their work done from home. Working from home, in a quiet and uninterrupted environment, can be an invaluable resource. There are good reasons to have offices but also good reasons to stay away from them periodically during the workweek. In subtle ways you can take advantage of working from home even without having a dedicated day at home. For example, one approach I've seen work very well for some is to get up a little early to work from home for an hour or two but then leave the office a little early in the afternoon to compensate for the extra time put in. Moreover, having done some very productive work in the morning can help alleviate the guilt of not staying late at the end of the day.

Now that you know more about the value of quiet, if you are able to arrange for a day or more a week working from home, perhaps this will convince you it is worth doing so. Even if it's not a dedicated space, merely a quiet one, it can be a great way to help set up the conditions for two awesome hours.

THE EFFECTS OF LIGHT ON PRODUCTIVITY

Noise is not the only environmental factor that deeply affects your productivity. Another stimulus, which you can often control, is light.

The reason light matters is because our eyes are not just for vision. In 2002, a discovery made at Brown and Johns Hopkins universities revolutionized how we think about our eyes. Up to that point only two kinds of cells in the retina had been shown to respond to light—rods and cones—which we rely on for vision. However, another cell type was demonstrated to respond to light as well, for a nonvisual purpose. These cells connect to a part of the brain responsible for maintaining circadian rhythms.[9] They respond especially well to light at the blue end of the visual spectrum[10]—like the light of the sky on a clear blue day.

Circadian rhythms guide our sleep, wake, eating, and energy cycles throughout the day. Light that activates these cells can help to reset the circadian cycles. It is still not fully understood why, but research is starting to map out the ways in which light affects cognition and emotions. Both blue light and bright white light seem to enhance a number of the mental faculties that can help us be highly effective. As we'll see, that kind of light influences things like alertness and concentration, and it can help us recharge after mental fatigue.

For example, there is research that suggests that white light enriched with blue may help you feel more alert and think more clearly. A group in the UK was curious to see what would happen to workers in a typical office environment when it was exposed to this bluish-white light—light that has the feel of the light from a clear blue sky. The researchers conducted their study at a company with two nearly identical floors where people engaged in very similar work. They were thus able to create a remarkably similar set of conditions where they could test different lighting types and their effect on employees' performance.

These researchers found that workers who were exposed to bluish-white light in the daytime were more likely to see improvements in self-reported alertness, concentration, clear thinking, performance, and sleep quality. The workers also experienced a decrease in fatigue in the evenings.[11]

Bluish-white light has also been shown to enhance self-control and the ability to mentally rotate objects—which is an ability important in engineering, design, and many of the scientific and technical fields. A research team in Italy built a room in which they could precisely control the lighting and even the ways that the light would meet someone's eyes as they sat at a desk.

All participants in their study were asked to perform executive-function tasks requiring self-control as well as take a mental rotation test. In the mental rotation test, they were shown a picture of a 3-D object and had to imagine turning it around to "see" whether it matched a picture of a different 3-D object in a different orientation. They all did these tests under white halogen lighting. Then the participants repeated the same kinds of tests, but one group did so under the warmer halogen lighting

again and a second group did so under cooler LED lighting (which has more of the blue part of the spectrum in
it). Both groups performed the same in the first series of
tests, working under white halogen lighting, but during
the second round of tests, those who worked under the
cooler LED lighting were more effective at executive control and more accurate at mental rotation.[12]

Bluish-white light, however, is not the only kind of
light that can be helpful with effectiveness. One group
in the Netherlands explored the consequences of bright
lights on alertness by having people wear a light meter on
their heads from eight A.M. to eight P.M. on three consec-
utive days, to measure how much light they were exposed
to during daylight hours. Every hour throughout the
day, participants were asked how they were feeling. The
research group found that good exposure to bright light
had an immediate effect on what they called "feelings of
vitality" and alertness—think of that as the opposite of
fatigue. The effects of bright lights were even greater earlier in the day and during the winter.[13]

Much as with noise, there seems to be a special set of
conditions for creativity, however. A study out of Germany
found that the opposite of bright lights—dim lighting—is
more conducive to creativity. Participants were arranged
in clusters in working environments with different light
levels: dim lighting, something approximating typical office
lighting, and very bright lighting. Then all participants were

asked to imagine they had just gone to a galaxy far away and to draw the alien life-form they encountered in their imagined planet. (It sounds like a fun study to participate in.) Those participants working in dim lighting generated more creative ideas than participants working in the other two lighting conditions. The researchers showed that the dimmer lighting led people to feel more free from constraints, which in turn led to more creativity. The key to the effects of light on creativity might not be the light level itself, therefore, but rather that dim lights give us the sense of being free from constraints. If that is the case, the authors of that research point out, windows can in principle give us that same feeling—the sense of being free from constraints—and thus lead to greater creativity.

This same study, however, showed that aside from having a positive effect on creativity, dim lighting took a toll on productivity. Participants who worked under dim lighting conditions did a little worse on analytical tasks that required focus and logical rule following.[14]

To make use of these lessons, there are several things you can do during those times when you want to be at your best:

- Turn on more lights. A brightly lit room is better for being at your mental best than a darker one, especially if it's a cloudy day or the middle of winter. If you have to, bring your own lamp to the office.

- If you can, be somewhere with ambient natural light on a day with clear blue skies, and set yourself up to work there.
- Consider replacing the current lightbulbs in your workspace with white lights that include more of the blue spectrum, even if it's just at your desk lamp. Research suggests there's a good chance you'll activate your eyes' retinal photoreceptive cells that communicate with your brain's circadian clock, helping you stay more alert.
- Dim your lights a bit or find a spot that's a little darker than usual when you want to work on a project that requires creativity.

SET UP YOUR IMMEDIATE WORKSPACE TO MAKE WORK EASIER

So far, we've explored how two factors of your workplace environment—noise and light—can have a great influence on your productivity. At first glance, those may seem like factors you have little control over, but you've now seen a number of ways that you can take enough control of both, at least for short periods.

Now we'll turn to the third part of the work environment that I believe we can influence in some useful ways: our immediate workspace. Our ability to set up our desks, cubicles, or offices also may not seem at first like something we have a huge amount of control over, but there are some important things we all can do that research suggests can make a difference in how we function psychologically.

In our immediate workspaces, four factors that matter to our productivity, and that we have some control over, are

- clutter,
- how constricted or expansive the space allows our movements to be,
- whether the space allows us to easily get up from our desks to move our bodies regularly, and
- whether there are objects and visuals in the space that help restore our mental energy.

When I think of clutter, the image that most frequently pops up is of a professor sitting at a desk strewn with papers and books dating back twenty years. Despite his overstuffed office and messy desk, he continues to be incredibly productive—a prolific published author and admired teacher. Whether he is productive because of or in spite of the clutter, we'll never know, because that professor will never clean up his mess.

Perhaps clutter works for a very few people. But for the vast majority of us, clutter is a hindrance to our mental performance. We're fooling ourselves if we think it doesn't matter enough to our productivity to do something about it. In strategy 3, we talked about how our brains are distraction-finding machines. More specifically, research shows that the greater the competition for our attention, the harder it is for us to exert control over what we focus on.[15] The clutter in your immediate workspace—the meeting notes, mystery papers, random borrowed books, product packages, training manuals, that extra phone headset you bought but never use, and so on—could be a major source of the distractions fighting for your attention.

Many of the things that may clutter your desk, in particular, are reminders of tasks on your to-do list: a note about a potential client you feel terrible about having taken so long to get back to, a document for a project that makes you stressed out to even think about. On top of just adding to the list of things competing for your attention, these are emotionally charged and somewhat threatening. When we detect something threatening, signals in the brain pertaining to that object or situation are amplified by activity from a part of the brain called the amygdala, in such a way that those threatening things are even more likely to grab our attention.[16]

So these kinds of physical reminders are especially bad to have haphazardly strewn about a place where you want to get important work done. Leaving things out on your desk as a way of remembering them will probably work to help you remember them. It is a strategy many people use.[17] But remembering them is precisely the problem when you are looking for a space in which to have two awesome hours of productivity.

So if you have to focus on a project or task, either clear away the clutter or find a clear space you can work in. We tend to pile up clutter wherever we go, so knowing how much extra mental energy you expend resisting distractions may help motivate you to clear your workspace when you want to get great work done.

If the idea of organizing and clearing the clutter is itself overwhelming for you, simply move all the clutter to another place (a file cabinet, a storage room, etc.) so that you don't see it when you sit down to work. Yes, that doesn't address the problem of the clutter itself and creates a different messy space. But this way you can be more productive when you need to be for your important work. Then find some unproductive time to go through that clutter later.

In addition to freeing us from distractions, how we set up our immediate workspace may actually influence how innovative we are and how much leadership presence or confidence we have. To be innovative requires we take risks to do things differently. It turns out that the ways we move within our workspace environment can affect our likelihood of taking the kind of leadership initiative and risks necessary for innovation.

Picture yourself sitting in a nice chair at your desk, with enough space to push back and put your feet up. You take on what psychologists call an expansive, open body posture that in the United States is iconic of being in charge. Professor Dana Carney (now at University of California at Berkeley's Haas School of Business) led a Columbia and Harvard University team of researchers who showed that taking on an expansive and open posture—what they called a "power pose," based on how power has historically been communicated nonverbally—for two minutes was enough to increase testosterone and decrease cortisol hormones, and it made people more comfortable with taking risks.[18] The combination of increased testosterone and decreased cortisol is a profile that has also been connected with being a more effective leader.[19] Being comfortable with taking risks and bringing to the table more leadership confidence can be incredible assets for the right kind of work challenge.

How you set up your workspace can actually facilitate and even encourage open, expansive postures or movements—those that prime risk taking and confident leadership, though risk taking comes in many forms. One study placed objects on a desk, either spread out where participants would need to reach for them—thus requiring them to make incidental expansive movements whenever they reached for those objects—or set tightly around a small space in front of the participants. The latter layout encouraged the typical small range of movements many office workers employ when sitting in front of their computers. When participants took on expansive and open postures—just as secondary consequences of how the work materials were laid out—the researchers saw increased tendencies to cheat while scoring themselves on word puzzles.[20] Cheating, of course, is not a desirable behavior, but it is a form of risk taking, so it fits with the other research on expansive postures.

The authors of this study make the case that increased risk taking is not always a plus, since there are circumstances when it can lead people to embrace the darker side of risk taking. But when you are engaged in a good task that requires you to be bold and take risks in order to do it well, I suggest you consider whether the space immediately around you allows you to move in ways that lead to expansive postures. Having cleared away the clutter, by the way, will make it easier to organize your workspace so as to encourage expansive movement.

Another challenge for the modern office worker, beyond the small, tight spaces in which we are frequently asked to work, is that we tend to sit relatively still in those positions for many hours of the day. As was discussed in the previous strategy, sometimes getting a bit of exercise is just what we need in order to be in top mental shape. Now a research group in the UK has found that sitting for even fifteen minutes can make us less energetic, and less positive emotionally.[21] This finding is, of course, troubling, if only because most professionals have to sit for long periods of time—typically in front of their computers—to do their jobs.

Today, though, we have more options available. Some

people have replaced their traditional chairs and desks with "treadmill desks"—essentially walking as they work. Does it increase productivity? A group of researchers at Stanford University found that walking on a treadmill or going for a walk outside increased creativity and improved people's facility with drawing analogies.[22] Some of the effects even carried over for a while once people were seated again.

Not everyone can replace their chair with a treadmill, but you can move more often and make adjustments to your workspace or select a workspace that makes it easier to move. Finding a place to stand and work, and another to sit and work, can give you the chance to trade off, moving and using your body differently each time you switch. There are also places you can work that make it easier or harder to get up and move around every fifteen minutes. For example, as much as I love coffee shops for reading a book, catching up with friends, or even a little creative work from time to time, when I set up there to work, it's very difficult to leave the seat (and my belongings) to walk around or move—certainly not every fifteen minutes.

Regardless of whether you have a clear desk or you move in ways that help you work more productively, at some point you will get fatigued. To help address that issue, you can set up your workspace with objects and visuals that will help restore your mental energy. A number of environmental factors have been shown to help people recharge: plants can help with attention capacity, both bird sounds and views of water are reported to be restorative, and a workspace that you've had a chance to personalize can help combat the feelings associated with emotional exhaustion that can result from working in an office without much privacy.[23]

Here are some things you can do to your workspace— that are within your control—to enhance your chances of being productive for a couple of hours:

- Clear the clutter. Do it late in the afternoon or evening when you don't have much mental energy left to engage in more productive work. If you don't have the time to clear it, simply move it somewhere that is out of sight.

- Place your phone, your glass of water, your pen, and any other work tools at the far corners of your desk, where you will need to reach for them expansively. If you feel tense, sit back for a minute, expand your chest, and spread your arms out. Maybe even buy a footstool or ottoman for your office on which you can put your feet up. Taking on these expansive "power poses" can shift your mental state.

- Don't sit at your desk for too long. We tend to become engrossed in working, so it will probably not be too much if you get up every time you think of doing so. If you can choose your workspace, choose one where getting up and moving around is easy to do.

- Personalize your workspace in some way. Specifically, consider adding some plants or images of water. When you personalize your space, though, don't do it by adding clutter to your desk.

How you set up your immediate area—whether you keep it cluttered or fairly clean, whether you allow enough space to make expansive postures or fit in a treadmill, or whether you keep an orchid or feature a painting of the ocean—can make an impact on whether that space is one where you can be super productive when you want to be.

WORK WITH YOUR SURROUNDINGS IN MIND

When we plan our two awesome hours, where we work matters. I've talked about three aspects of our workspaces that make a difference and that we can have some control over. The first is how much noise we expose ourselves to. The second is whether the lighting is bright and cool colored, rather than dim and warm colored. And the third is how we set up our immediate workspace—whether our desks are clear or cluttered, whether we can move expansively, whether our ability to get up and move around is hindered or enabled, and whether the space includes some restorative elements. Let's take a look at how Samantha, the CFO of a start-up we met earlier, changed her environment to ensure she is as productive as she needs to be in her new role.

Samantha, as you may recall, faces the same challenges most of us do: she cannot change her company's office layout or change how little privacy she has. So she found a compromise: several times a week, when she needs her brain firing on all cylinders, she reserves one of the small conference rooms in her office for a couple of hours. There, sitting at a spacious conference table where she can spread out, free from noisy distractions, she is able to finally focus.

To be more productive when working at her desk, she cleared away the piles of papers that had accumulated since she started at the company. Her first six months at the company had been tough and went by in a flash, so she hadn't taken the time to personalize her workspace. She brought in pictures of her family and a print she bought while on vacation in Hawaii, featuring a serene ocean scene. She also now uses some noise-cancelling headphones to drown out the noise around her when she needs to focus at her desk—but she never listens to music except when she is working on routine tasks.

Although Samantha's workspace and her company's open floor plan isn't great for doing work that requires intense concentration, it does facilitate collaboration. By working with her surroundings to increase her ability to focus uninterrupted, she is getting the best of both worlds: the ability to retreat and focus on her individual work when she needs to and an office where she can work closely with her colleagues.

I've saved this strategy on how your work environment affects your productivity for last because it can serve as a tool or a hindrance for all of the other strategies in the book. Leveraging your decision points throughout the day, strategically ordering your tasks so as to manage your mental energy, removing distractions and learning to mind wander more effectively, and exercising and eating to increase your productivity can all be facilitated by your surroundings.

The space in which you work influences how often you will be interrupted, which leads to your needing a plan for how to handle the decision points those interruptions will create. Where you work also plays into how much mental energy you will expend on self-control and what emotional triggers may occur each day. Apart from interruptions, your workspace can be booby-trapped with potential distractors or relatively free of them. Whether that workspace has or is near a place to go for a rigorous walk or to exercise can make a difference in whether you find ways to incorporate exercise too.

So spend some time today looking around you. Is your work environment set up to help you achieve two awesome hours of productivity?

CONCLUSION

I've never met a person who didn't lament the fact that they can't seem to get enough done. We're all overwhelmed with work and life demands. And up until now, most people have responded to this by turning their attention to how to be more efficient. *How can I get rid of unnecessary downtime? How can I get my people to put in more hours on work each week? How can I move from task to task—or even multitask—so there's not even one wasted moment?*

It turns out we've been barking up the wrong tree. Efficiency is a metric for machines and computers. But science is revealing that humans are not just computers on life support. We have brains and bodies, and we operate according to biological needs. The right metric for human performance is effectiveness, not efficiency. Our brains can be remarkably effective under the right conditions and not so effective under the wrong ones. Neuroscience and psychology reveal what those conditions are and how we can set ourselves up for highly effective mental performance.

After you read this book, rather than structuring your time around how to cram more work hours into a day, I hope you'll find yourself structuring your day around how to have two awesome hours. From food to exercise to time of day to types of cognitive tasks, I hope you will plan your activities and your environment in ways that help you achieve a couple of hours of real effectiveness, when you can get the important work done. You can start using each strategy offered in this book right away. The strategies help you cultivate those effective hours.

STRATEGY 1: RECOGNIZE YOUR DECISION POINTS

The moment when you have finished a task and are free to start the next one is a precious opportunity. As simple as it may appear, learning to recognize and use your few daily decision points is a game changer.

You may often spend hours stuck on unimportant work without thinking. Once you engage in a task, your brain tends to switch to autopilot, and as a result, you can easily continue doing the same task until it ends or something interrupts you. If you are answering an e-mail, your brain gets into e-mail mode, and you can get lost answering one after another. So while you may have intended to only answer e-mails for one hour, before moving on to that very important work you absolutely had to get to before lunch, three hours later, you can find yourself saying, *Okay, I'll respond to this one more e-mail and then I'll turn to that important task.*

When your brain goes into automatic mode, you become less aware of your surroundings and the time passing. That's why the decision points in your day are so useful: they are the moments when you snap out of automatic and become aware that you can decide how to spend your time. Most of the time, you probably rush through them, allowing your brain to grab on to whatever seems most expedient or urgent to it at the time. However, you can step back and thoughtfully consider what is worth doing and what is worth ignoring.

You can learn to recognize these decision moments, savor them, and stick with them for a few minutes until you are able to reconnect with your real priorities. You can use them to figure out what makes the most sense to do given the time on your hands and your mental energy. There's a time and place for the less important work, but leveraging your decision points will help you keep attuned to your larger priorities.

STRATEGY 2:
MANAGE YOUR MENTAL ENERGY

Managing time is not just about scheduling. Not all hours are the same. Because the brain fatigues and needs rest to regenerate, the best way to get your work done is not to find the hours to do it on a spreadsheet or calendar. Rather, the key is to tackle your work when you have the right mental energy for it.

Many tasks you do take a toll on you and your brain. Some deplete your mental energy, while others elicit strong emotions. This is not a bad thing—it is just how your mind reacts to the world around you. But it is worth paying close attention to so you can make the best use of your mental energy. Armed with knowledge of the kind of tasks that are most likely to lead to mental fatigue and how to anticipate the emotions that are likely to surface throughout the day, you can make wise decisions. You can choose when to work on what and when to take a moment to regulate your emotions.

Wearing yourself out mentally with a series of un-important tasks that tax your executive functions right before you need to bring your A game to a situation, like a presentation or important meeting, is a terrible idea. But you can do two things: schedule your tasks to take advantage of when your mental energy is at its peak and strategically choose not to do something that will drain you. Managing your mental energy like this will set you up for a couple of hours of true productivity.

STRATEGY 3: STOP FIGHTING DISTRACTIONS

As critical as we claim attention is for our productivity, we sure have mistaken ideas about how it operates. For instance, we often describe it as a spotlight. But a spotlight stays put once you shine it on a spot. Our neural mechanisms for attention do not work that way at all. Our attention systems are designed to regularly refresh—to be ready to discover what is new in the environment and to help us navigate a constantly changing world. No wonder when you force yourself to stay focused you become maddeningly frustrated as you find your attention drifting.

But letting your mind wander may be exactly what you need to do in order to find your focus again. Getting back on track should happen far quicker that way than if you jump to whatever other task grabs your attention. Reading the sports page or checking social media sites will often distract you for a half hour or longer, whereas staring out the window until you snap out of your daydream will probably have you back and focusing on your task in minutes, having engaged in some important cognitive processes that occur while your mind wanders.

STRATEGY 4: LEVERAGE YOUR MIND–BODY CONNECTION

The feeling of being overwhelmed is an emotion, and emotions are not just "in your head" but are highly physical. The awareness that you have more to do than you can cope with is usually accompanied by physical sensations like a knot in your stomach. That's because your body and your mind are intricately intertwined.

It is less surprising from this vantage point that exercising can be such a reliable way to reduce anxiety, elevate your mood, and help with your cognitive functions. You don't have to overdo it; in fact, it's better not to work out too hard when it's for immediate mental benefits. Moderate exercise can be ideal.

Likewise, you can eat, stay hydrated, and drink caffeinated beverages in ways that help set up the conditions for your being highly effective when it's important. Smaller meals consumed more often, a regular intake of water, and no more than your normal dose of caffeine can facilitate your being at your best.

How you treat your body has real consequences on how your mind performs. Skip exercise and eat whatever you want on your downtime if you like. But when it's time to be on, using your body and eating in ways that set you up for success in the hours that follow can separate you from the pack.

STRATEGY 5: MAKE YOUR WORKSPACE WORK FOR YOU

The final strategy for creating at least two awesome hours of productivity is to work in the right physical environment for your brain to be highly productive. The environment where you work affects what you accomplish in more ways than you might expect—and thus the likelihood that you will use your time effectively. Noise makes it hard to focus, and you can expect plenty of it in this age of cubicles and open floor plan offices. Working under different kinds of lights can make a big difference to your mental alertness and creativity. And the immediate workspace can be either restorative or distracting, easy or difficult to move in, and even inspire you to take risks, depending on how you organize the space and your things within it. You often can't change the place where you work, but there are lots of little things you can do to ensure that your workspace is helping, not hindering, your productivity.

—

THESE STRATEGIES ARE effective not only because they are simple and easy to start implementing, but also because they work with, not against, your biology. Technology will continue to make it possible for us to pack more into our days. Our work culture will likely continue to push us to be more efficient and give our all at every step. And we will likely continue to have more and more reasons to feel overwhelmed by the extensive demands placed on us. As that pattern accelerates, it becomes ever more important, I believe, to understand how human beings work best. This understanding will enable us to adapt and succeed in this incredibly rigorous environment.

The strategies I share in this book address what I believe is the biggest challenge posed by our current work culture: being overwhelmed. By becoming students of how human beings can work most effectively, we all can increase our self-compassion, master our work, and gain control over our lives.

ACKNOWLEDGMENTS

This book is the result of the tremendous guidance and support of a remarkable team of people. Heidi Grant Halvorson planted the seed that grew into this book when she introduced me to my agent, Giles Anderson. Giles walked me through every step from idea creation to finding the right publisher and inspired me to believe in myself and to keep moving fast, along the way. My editor, Genoveva Llosa, has with a steady hand and a keen eye captured what I wanted to say and helped me communicate it more effectively. She has shown me how to sculpt these messages to reach our audience in meaningful and applicable ways. Along with Genoveva, Hannah Rivera, her editorial assistant, Noël Chrisman, my production coordinator, and Dianna Stirpe, my copyeditor, helped me produce a far better manuscript than I could ever have done on my own. theBookDesigners created a cover that powerfully captures the message and promise of the book in an instant. Jennifer Jensen, my marketer, Suzanne Wickham, my publicist, and the rest of the marketing, PR, executive, international, translation, production, and audio teams at HarperOne and HarperCollins believed in the potential of this book, and championed it. David and Lisa Rock significantly deepened my understanding of the business world as well as provided valuable encouragement. Steve Leeds and Rachel Hott provided generous mentorship and key opportunities for me to develop my ideas. Jenny Xiao and Peter Mende-Siedlecki provided exceptional assistance in verifying scientific accuracy and fact checking.

I am grateful to all of my friends, relatives, colleagues, and work relationships who shared their enthusiasm when I discussed the book, giving me more energy to work. My brother, Kenny Davis, guided me to invaluable insights about messaging. My parents, Susan and Don Davis, helped me tremendously as I developed the conceptual frameworks. And my wife, Daniela, was the engine that drove me. At every stage, she selflessly gave me what she saw that I needed—feedback, motivation, freedom, and compassion—to make it not only easy but exciting to work on the book.

I have been blessed with an extraordinary team who have made it possible to write a book that I believe can make a meaningful difference to many people in how we handle being overwhelmed, and how we understand ourselves.

NOTES

INTRODUCTION: BE AWESOMELY EFFECTIVE

1. Benjamin Franklin, "Autobiography of Benjamin Franklin," public domain (published January 1, 1790), https://itun.es/us/xZiNx.l.

2. Benjamin Franklin, "I Sing My Plain Country Joan, 1742," Founders Online, National Archives, last modified December 1, 2014, http://founders.archives.gov/documents/Franklin/01-02-02-0087.

3. "Benjamin Franklin," Biography.com, www.biography.com/people/benjamin-franklin-9301234.

4. For a more complete definition of embodied cognition, see this introduction to a special issue of a journal on the subject that I coedited with Art Markman, Ph.D.: Joshua I. Davis and Arthur B. Markman, "Embodied Cognition as a Practical Paradigm: Introduction to the Topic, the Future of Embodied Cognition," *Topics in Cognitive Science* 4, no. 4 (2012): 685–91. For a curated collection of papers and other resources on embodied cognition, visit www.embodiedmind.org.

5. Dana R. Carney, Amy J. Cuddy, and Andy J. Yap, "Power Posing: Brief Nonverbal Displays Affect Neuroendocrine Levels and Risk Tolerance," *Psychological Science* 21, no. 10 (2010): 1363–68; Pranjal H. Mehta and R. A. Josephs, "Testosterone and Cortisol Jointly Regulate Dominance: Evidence

for a Dual-Hormone Hypothesis," *Hormones and Behavior* 58 (2010): 898–906.

6. Jesse Chandler and Norbert Schwarz, "How Extending Your Middle Finger Affects Your Perception of Others: Learned Movements Influence Concept Accessibility," *Journal of Experimental Social Psychology* 45, no. 1 (January 2009): 123–28.

7. Josh Davis, Maite Balda, David Rock, Paul McGinniss, and Lila Davachi, "The Science of Making Learning Stick: An Update to the AGES Model," *NeuroLeadership Journal* 5 (2014). Note that this example about learning is not one that all embodied cognition researchers would include as embodied cognition. However, I deliberately place it here to help the reader see that these kinds of neuroscience lessons illustrate consequences of having a body that operates unlike a typical computer or nonhuman machine.

STRATEGY 1: RECOGNIZE YOUR DECISION POINTS

1. Charles Duhigg, *The Power of Habit: Why We Do What We Do and How to Change* (New York: Random House, 2013).

2. Susan T. Fiske and Shelley E. Taylor, *Social Cognition: From Brains to Culture* (Thousand Oaks, CA: Sage Publications, 2013).

3. "Trance," Merriam-Webster online, retrieved September 14, 2014, www.merriam-webster.com/dictionary/trance.

4. Ezequiel Morsella, "The Function of Phenomenal States: Supramodular Interaction Theory," *Psychological Review* 112, no. 4 (2005): 1000–21. This work was originally published while Dr. Morsella was at Yale University. He then established his lab group at San Francisco State University.

5. John G. Kerns et al., "Anterior Cingulate Conflict Monitoring and Adjustments in Control," *Science* 303, no. 5660 (2004): 1023–26; and Matthew M. Botvinick, Jonathan D. Cohen, and Cameron S. Carter, "Conflict Monitoring and Anterior

Cingulate Cortex: An Update," *Trends in Cognitive Sciences* 8, no. 12 (2004): 539–46.

6. Naomi I. Eisenberger and Matthew D. Lieberman, "Why Rejection Hurts: A Common Neural Alarm System for Physical and Social Pain," *Trends in Cognitive Sciences* 8, no. 7 (2004): 294–300.

7. Kathleen D. Vohs and Brandon J. Schmeichel, "Self-Regulation and Extended Now: Controlling the Self Alters the Subjective Experience of Time," *Journal of Personality and Social Psychology* 85, no. 2 (2003): 217–30.

8. Stephen Covey, *The Seven Habits of Highly Effective People* (New York: Free Press, 1989).

9. Something Benjamin Franklin aimed to do every morning, in order to reconnect with what was important to him, was to ask himself, "What good shall I do this day?" I don't know if he took a moment to think about that later in the day, but it strikes me as a similar idea. Excerpt from Benjamin Franklin, "Autobiography of Benjamin Franklin," public domain (published January 1, 1790), https://itun.es/us/xZiNx.l.

10. Nira Liberman and Yaacov Trope, "The Psychology of Transcending the Here and Now," *Science* 322, no. 5905 (2008): 1201–5.

11. Gal Zauberman et al., "Discounting Time and Time Discounting: Subjective Time Perception and Intertemporal Preferences," *Journal of Marketing Research* 46, no. 4 (2009): 543–56.

12. Aleksandra Luszczynska, Anna Sobczyk, and Charles Abraham, "Planning to Lose Weight: Randomized Controlled Trial of an Implementation Intention Prompt to Enhance Weight Reduction Among Overweight and Obese Women," *Health Psychology* 26, no. 4 (2007): 507–12.

13. Thomas L. Webb et al., "Effective Regulation of Affect: An

Action Control Perspective on Emotion Regulation," *European Review of Social Psychology* 23, no. 1 (2012): 143–86.

14. Janine Chapman, Christopher J. Armitage, and Paul Norman, "Comparing Implementation Intention Interventions in Relation to Young Adults' Intake of Fruit and Vegetables," *Psychology and Health* 24, no. 3 (2009): 317–32.

15. Peter M. Gollwitzer, "Implementation Intentions: Strong Effects of Simple Plans," *American Psychologist* 54, no. 7 (1999): 493–503; and Peter M. Gollwitzer and Paschal Sheeran, "Implementation Intentions and Goal Achievement: A Meta-Analysis of Effects and Processes," *Advances in Experimental Social Psychology* 38 (2006): 69–119.

16. Michael L. Anderson, "Neural Reuse: A Fundamental Organizational Principle of the Brain," *Behavioral and Brain Sciences* 33, no. 04 (2010): 245–66; and Michael L. Anderson, Michael J. Richardson, and Anthony Chemero, "Eroding the Boundaries of Cognition: Implications of Embodiment," *Topics in Cognitive Science* 4, no. 4 (2012): 717–30.

17. M. Brouziyne and C. Molinaro, "Mental Imagery Combined with Physical Practice of Approach Shots for Golf Beginners," *Perceptual and Motor Skills* 101, no. 1 (2005): 203–11.

18. Sonal Arora et al., "Mental Practice Enhances Surgical Technical Skills: A Randomized Controlled Study," *Annals of Surgery* 253, no. 2 (2011): 265–70.

19. Mike Knudstrup, Sharon L. Segrest, and Amy E. Hurley, "The Use of Mental Imagery in the Simulated Employment Interview Situation," *Journal of Managerial Psychology* 18, no. 6 (2003): 573–91.

20. Vinoth K. Ranganathan et al., "From Mental Power to Muscle Power—Gaining Strength by Using the Mind," *Neuropsychologia* 42, no. 7 (2004): 944–56.

STRATEGY 2: MANAGE YOUR MENTAL ENERGY

1. Elliot T. Berkman and Jordan S. Miller-Ziegler, "Imaging Depletion: fMRI Provides New Insights into the Processes Underlying Ego Depletion," *Social Cognitive and Affective Neuroscience* 8, no. 4 (2012): 359–61; and Michael Inzlicht and Brandon J. Schmeichel, "What Is Ego Depletion? Toward a Mechanistic Revision of the Resource Model of Self-Control," *Perspectives on Psychological Science* 7, no. 5 (2012): 450–63.

2. Jessica R. Cohen and Matthew D. Lieberman, "The Common Neural Basis of Exerting Self-Control in Multiple Domains," in *Self Control in Society, Mind, and Brain,* ed. Ran R. Hassin, Kevin N. Ochsner, and Yaacov Trope (New York: Oxford University Press, 2010), 141–60; Matthew D. Lieberman, "The Brain's Braking System (and How to 'Use Your Words' to Tap into It)," *NeuroLeadership Journal* 2 (2009): 9–14; Elliot T. Berkman, Lisa Burklund, and Matthew D. Lieberman, "Inhibitory Spillover: Intentional Motor Inhibition Produces Incidental Limbic Inhibition via Right Inferior Frontal Cortex," *Neuroimage* 47, no. 2 (2009): 705–12; and Michael Inzlicht, Elliot Berkman, and Nathaniel Elkins-Brown, "The Neuroscience of 'Ego Depletion' or: How the Brain Can Help Us Understand Why Self-Control Seems Limited," in *Social Neuroscience: Biological Approaches to Social Psychology,* ed. Eddie Harmon-Jones and Michael Inzlicht (New York: Psychology Press, 2015). Inzlicht, Berkman, and Elkins-Brown point out that there are several brain regions necessary for self-control in different capacities (e.g., the ventrolateral prefrontal cortex is especially important for inhibition). Readers interested in the neuroscience of self-control are encouraged to read that chapter of *Social Neuroscience* for more detail.

3. Inzlicht, Berkman, and Elkins-Brown (see note 2 of this strategy) argue that the evidence is best understood to mean that self-control takes effort, and after a while of it, the vari-

ous brain regions involved engage in less of it because we lose the motivation to keep at the effortful activity. However, they review research showing that with sufficient motivation— e.g., rewards of various kinds or the belief that we can sustain ourselves—our brains are capable of persisting at self-control when we need to for a very long time. The exact amount of time is unknown.

4. As an alternative to using willpower to say no (e.g., to a donut), you can avoid considering the donut in the first place, so saying no doesn't require any self-control. Here's one way to achieve something like this: Before thinking about what to eat, reframe what it means to eat breakfast as something you do to make yourself feel good for the whole morning, all the way until lunch. Donuts, on the other hand, will make you feel good for only a few minutes and tired or hungry a couple of hours later. That reframing may make it so that you both avoid the donut and don't need to use up any of your self-control, because donuts are simply not appealing from that vantage point. I talk more about this way of thinking about food in strategy 4.

5. Roy F. Baumeister and John Tierney, *Willpower: Rediscovering the Greatest Human Strength* (New York: Penguin, 2011), 99.

6. Kathleen D. Vohs et al., "Making Choices Impairs Subsequent Self-Control: A Limited-Resource Account of Decision Making, Self-Regulation, and Active Initiative," *Journal of Personality and Social Psychology* 94, no. 5 (2008): 883–98.

7. Shaheem Reid, with additional reporting by Sway Calloway, "All Eyes on Beyoncé," MTV.com, www.mtv.com/bands/b/beyonce/news_feature_081406/.

8. Eddie Harmon-Jones et al., "The Effect of Personal Relevance and Approach-Related Action Expectation on Relative Left Frontal Cortical Activity," *Psychological Science* 17, no. 5 (2006): 434–40.

9. Each of these findings regarding sadness is reviewed in Joseph P. Forgas, "Don't Worry, Be Sad! On the Cognitive, Motivational, and Interpersonal Benefits of Negative Mood," *Current Directions in Psychological Science* 22, no. 3 (2013): 225–32.

10. Matthijs Baas, Carsten K. W. De Dreu, and Bernard A. Nijstad, "A Meta-Analysis of 25 Years of Mood-Creativity Research: Hedonic Tone, Activation, or Regulatory Focus?" *Psychological Bulletin* 134, no. 6 (2008): 779–806.

11. Maya Tamir, "Don't Worry, Be Happy? Neuroticism, Trait-Consistent Affect Regulation, and Performance," *Journal of Personality and Social Psychology* 89, no. 3 (2005): 449–61. This research suggests that these effects were most pronounced for the more neurotic people in the study's sample.

12. Karuna Subramaniam et al., "A Brain Mechanism for Facilitation of Insight by Positive Affect," *Journal of Cognitive Neuroscience* 21, no. 3 (2009): 415–32.

13. Baas, De Dreu, and Nijstad, "A Meta-Analysis of 25 Years," 779–806; and Alice M. Isen, Kimberly A. Daubman, and Gary P. Nowicki, "Positive Affect Facilitates Creative Problem Solving," *Journal of Personality and Social Psychology* 52, no. 6 (1987): 1122–31. In particular, in the Baas, De Dreu, and Nijstad article it appears that positive emotions that are approach oriented and relatively more active help with creativity. Other positive emotions, like feeling calm or relaxed, might not.

14. Suzanne K. Vosburg, "The Effects of Positive and Negative Mood on Divergent-Thinking Performance," *Creativity Research Journal* 11, no. 2 (1998): 165–72; and Norbert Schwarz, Herbert Bless, and Gerd Bohner, "Mood and Persuasion: Affective States Influence the Processing of Persuasive Communications," *Advances in Experimental Social Psychology* 24 (1991): 161–99. Positive mood can lead to "satisficing"—to being comfortable with a decision, solution, or answer that is good enough, rather than pressing on for what might be ideal.

Interestingly, in light of the other findings cited earlier that positive mood enhances creativity, Vosburg argues that there are at least some special circumstances in which positive mood might not enhance creativity, such as when demanding an ideal or optimal creative solution.

15. Guido Hertel et al., "Mood Effects on Cooperation in Small Groups: Does Positive Mood Simply Lead to More Cooperation?" *Cognition and Emotion* 14, no. 4 (2000): 441–72.

16. Joseph P. Forgas, "On Feeling Good and Getting Your Way: Mood Effects on Negotiator Cognition and Bargaining Strategies," *Journal of Personality and Social Psychology* 74, no. 3 (1998): 565–77.

17. Martin E. P. Seligman and Mihaly Csikszentmihalyi, "Positive Psychology: An Introduction," *American Psychologist* 55, no. 1 (2000): 5–14. When reading the research, it's harder to get a handle on specifically which positive emotions have different effects than it is with the negative emotions. Some studies may specify happiness, amusement, satisfaction, comfort, pride, and so on, but many simply describe the positive emotions more generally. My guess is that this is because more energy has been invested in studying what is negative rather than what is positive in psychology. Regardless of the reason, I believe it is more prudent at this point just to say that positive emotion—something like happiness, joy, amusement, or generally feeling good—has these consequences.

18. Some researchers distinguish between emotions—being shorter in duration and tied to a clear cause—and mood—being longer in duration and not as clearly tied to a specific cause. I am using the terms interchangeably because regardless of the duration or cause of the emotion or mood, the consequences I've described should be similar to the best of my knowledge.

19. Martin P. Paulus, "The Breathing Conundrum—Interoceptive Sensitivity and Anxiety," *Depression and Anxiety* 30, no. 4

(2013): 315–20; and A. D. Craig, "Interoception: The Sense of the Physiological Condition of the Body," *Current Opinion in Neurobiology* 13, no. 4 (2003): 500–5.

20. F. A. Bainbridge, "The Relation Between Respiration and the Pulse-Rate," *Journal of Physiology* 54, no. 3 (1920): 192–202.

21. Dianne M. Tice et al., "Restoring the Self: Positive Affect Helps Improve Self-Regulation Following Ego Depletion," *Journal of Experimental Social Psychology* 43, no. 3 (2007): 379–84.

22. Amber Brooks and Leon Lack, "A Brief Afternoon Nap Following Nocturnal Sleep Restriction: Which Nap Duration Is Most Recuperative?" *Sleep* 29, no. 6 (2006): 831–40. This research studied people napping after some sleep deprivation. You may not always be sleep deprived at work, but if you are, a nap may help you more than you think.

STRATEGY 3: STOP FIGHTING DISTRACTIONS

1. M. I. Posner, "Attention: The Mechanisms of Consciousness," *Proceedings of the National Academy of Sciences* 91, no. 16 (1994): 7398–403.

2. Research on habituation, as reviewed in Christian Balkenius, "Attention, Habituation, and Conditioning: Toward a Computational Model," *Cognitive Science Quarterly* 1, no. 2 (2000): 171–204.

3. Some readers may be wondering about how we can get lost in a good book or video game, attending for a very long time. Presumably, what is happening is that the story or adventure has novelty or suspense to grab attention nearly constantly. So the authors may be leveraging the fact that we have an attention system that delights in finding distractors.

4. Daniel M. Wegner et al., "Paradoxical Effects of Thought Suppression," *Journal of Personality and Social Psychology* 53, no. 1 (1987): 5–13. The original research on this topic used the white bear as the primary example, so I have used the polar bear here in honor of that.

5. Allan M. Collins and Elizabeth F. Loftus, "A Spreading-Activation Theory of Semantic Processing," *Psychological Review* 82, no. 6 (1975): 407–28.

6. A search through the scientific literature doesn't provide many answers about how long most people can attend to one thing. It is unclear why. It may be too dependent on context.

7. Elizabeth R. Valentine and Philip L. G. Sweet, "Meditation and Attention: A Comparison of the Effects of Concentrative and Mindfulness Meditation on Sustained Attention," *Mental Health, Religion, and Culture* 2, no. 1 (1999): 59–70. In that same study, among meditators, mindfulness practitioners had better sustained attention than concentrative meditators.

8. Jonathan Smallwood and Jessica Andrews-Hanna, "Not All Minds That Wander Are Lost: The Importance of a Balanced Perspective on the Mind-Wandering State," *Frontiers in Psychology* 4 (2013): 441.

9. Benjamin Baird et al., "Inspired by Distraction: Mind Wandering Facilitates Creative Incubation," *Psychological Science* 23, no. 10 (2012): 1117–22.

10. Benjamin Baird, Jonathan Smallwood, and Jonathan W. Schooler, "Back to the Future: Autobiographical Planning and the Functionality of Mind-Wandering," *Consciousness and Cognition* 20, no. 4 (2011): 1604–11.

11. Jonathan Smallwood, Louise Nind, and Rory C. O'Connor, "When Is Your Head At? An Exploration of the Factors Associated with the Temporal Focus of the Wandering Mind," *Consciousness and Cognition* 18, no. 1 (2009): 118–25.

12. Jon Kabat-Zinn, "An Outpatient Program in Behavioral Medicine for Chronic Pain Patients Based on the Practice of Mindfulness Meditation: Theoretical Considerations and Preliminary Results," *General Hospital Psychiatry* 4, no. 1 (1982): 33–47; and Jon Kabat-Zinn, *Full Catastrophe Living (Revised Edition): Using the Wisdom of Your Body and Mind to Face Stress, Pain, and Illness* (New York: Random House, 2013).

13. Maryanna D. Klatt, Janet Buckworth, and William B. Malarkey, "Effects of Low-Dose Mindfulness-Based Stress Reduction (MBSR-LD) on Working Adults," *Health, Education, and Behavior* 36, no. 3 (2009): 601–14.

14. Philippe R. Goldin and James J. Gross, "Effects of Mindfulness-Based Stress Reduction (MBSR) on Emotion Regulation in Social Anxiety Disorder," *Emotion* 10, no. 1 (2010): 83–91.

15. Linda E. Carlson and Sheila N. Garland, "Impact of Mindfulness-Based Stress Reduction (MBSR) on Sleep, Mood, Stress, and Fatigue Symptoms in Cancer Outpatients," *International Journal of Behavioral Medicine* 12, no. 4 (2005): 278–85.

16. Scott R. Bishop et al., "Mindfulness: A Proposed Operational Definition," *Clinical Psychology: Science and Practice* 11, no. 3 (2004): 230–41. Bishop et al. offer this description of mindfulness: "The first component involves the self-regulation of attention so that it is maintained on immediate experience, thereby allowing for increased recognition of mental events in the present moment. The second component involves adopting a particular orientation toward one's experiences in the present moment, an orientation that is characterized by curiosity, openness, and acceptance" (p. 232).

STRATEGY 4: LEVERAGE YOUR MIND–BODY CONNECTION

1. Andy Clark, *Being There: Putting Brain, Body, and World Together Again* (Cambridge, MA: MIT Press, 1997); Antonio Damasio, *Descartes' Error: Emotion, Reason, and the Human Brain* (New York: HarperCollins, 1994); and George Lakoff and Mark Johnson, *Philosophy in the Flesh: The Embodied Mind and Its Challenge to Western Thought* (New York: Basic Books, 1999). While it may fit with intuition for many people these days, in philosophical circles it is perhaps more revo-

lutionary to suggest that the mind and body are inextricably linked. Reading work on the topic makes it clear that most of us probably do have a large number of unchecked assumptions, reflecting a belief that our minds and bodies operate fairly separately. I believe such assumptions have consequences on what we think is worth doing—such as exercising.

2. Stanley Schachter and Jerome E. Singer, "Cognitive, Social, and Physiological Determinants of Emotional State," *Psychological Review* 69, no. 5 (1962): 379–99.

3. Nelson Mandela, *Long Walk to Freedom: The Autobiography of Nelson Mandela* (New York: Little, Brown and Company, 1994), 490.

4. Lot Verburgh et al., "Physical Exercise and Executive Functions in Preadolescent Children, Adolescents, and Young Adults: A Meta-Analysis," *British Journal of Sports Medicine* 48, no. 12 (2014): 973–79.

5. Shannan E. Gormley et al., "Effect of Intensity of Aerobic Training on VO2max," *Medicine and Science in Sports and Exercise* 40, no. 7 (2008): 1336–43.

6. Hiroki Yanagisawa, Ippeita Dan, Daisuke Tsuzuki, Morimasa Kato, Masako Okamoto, Yasushi Kyutoku, and Hideaki Soya, "Acute Moderate Exercise Elicits Increased Dorsolateral Prefrontal Activation and Improves Cognitive Performance with Stroop Test," *Neuroimage* 50, no. 4 (2010): 1702-10.

7. Kevin C. O'Leary et al., "The Effects of Single Bouts of Aerobic Exercise, Exergaming, and Videogame Play on Cognitive Control," *Clinical Neurophysiology* 122, no. 8 (2011): 1518–25.

8. Berit Inkster and Brian M. Frier, "The Effects of Acute Hypoglycaemia on Cognitive Function in Type 1 Diabetes," *British Journal of Diabetes and Vascular Disease* 12, no. 5 (2012): 221–26.

9. Franciele R. Figueira et al., "Aerobic and Combined Exercise Sessions Reduce Glucose Variability in Type 2 Diabetes:

Crossover Randomized Trial," *PLoS ONE* 8, no. 3 (2013):
e57733.

10. Steven J. Petruzzello et al., "A Meta-Analysis on the Anxiety-
Reducing Effects of Acute and Chronic Exercise: Outcomes
and Mechanisms," *Sports Medicine* 11, no. 3 (1991): 143–82.

11. Eli Puterman et al., "The Power of Exercise: Buffering the
Effect of Chronic Stress on Telomere Length," *PLoS ONE* 5,
no. 5 (2010): e10837.

12. Justy Reed and Deniz S. Ones, "The Effect of Acute Aerobic Exercise on Positive Activated Affect: A Meta-Analysis,"
Psychology of Sport and Exercise 7, no. 5 (2006): 477–514.

13. Reed and Ones, "The Effect of Acute Aerobic Exercise,"
477–514.

14. Bryan D. Loy, Patrick J. O'Connor, and Rodney K. Dish-
man, "The Effect of a Single Bout of Exercise on Energy and
Fatigue States: A Systematic Review and Meta-Analysis,"
Fatigue: Biomedicine, Health and Behavior 1, no. 4 (2013):
223–42.

15. Alexa Hoyland, Clare L. Lawton, and Louise Dye, "Acute
Effects of Macronutrient Manipulations on Cognitive Test
Performance in Healthy Young Adults: A Systematic Research
Review," *Neuroscience and Biobehavioral Reviews* 32, no. 1
(2008): 72–85.

16. Edward Leigh Gibson, "Effects of Energy and Macronutrient
Intake on Cognitive Function Through the Lifespan," *Pro-
ceedings of the Latvian Academy of Sciences, Section B, Natural,
Exact, and Applied Sciences* 67, nos. 4–5 (2013): 303–447.

17. Karina Fischer et al., "Cognitive Performance and Its Rela-
tionship with Postprandial Metabolic Changes After Ingestion
of Different Macronutrients in the Morning," *British Journal
of Nutrition* 85, no. 03 (2001): 393–405.

18. Gibson, "Effects of Energy and Macronutrient Intake,"
303–447.

19. David Benton, "Carbohydrates and the Cognitive Performance of Children," Carbohydrate News, Canadian Sugar Institute Nutrition Information Service, 2012, www.sugar.ca/SUGAR/media/Sugar-Main/PDFs/CarbNews2012_ENG-qxp_FINAL.pdf.

20. Hayley Young and David Benton, "The Glycemic Load of Meals, Cognition, and Mood in Middle and Older Aged Adults with Differences in Glucose Tolerance: A Randomized Trial," *e-SPEN Journal* 9, no. 4 (2014): e147–54.

21. Simon B. Cooper et al., "Breakfast Glycaemic Index and Cognitive Function in Adolescent School Children," *British Journal of Nutrition* 107, no. 12 (2012): 1823–32.

22. Paul Hewlett, Andrew Smith, and Eva Lucas, "Grazing, Cognitive Performance, and Mood," *Appetite* 52, no. 1 (2009): 245–48.

23. Dale A. Schoeller, "Changes in Total Body Water with Age," *American Journal of Clinical Nutrition* 50, no. 5 (November 1, 1989): 1176–81.

24. Ana Adan, "Cognitive Performance and Dehydration," *Journal of the American College of Nutrition* 31, no. 2 (2012): 71–78.

25. Lawrence E. Armstrong et al., "Mild Dehydration Affects Mood in Healthy Young Women," *Journal of Nutrition* 142, no. 2 (2012): 382–88.

26. Natalie A. Masento et al., "Effects of Hydration Status on Cognitive Performance and Mood," *British Journal of Nutrition* 111, no. 10 (2014): 1841–52.

27. Melanie A. Heckman, Jorge Weil, and Elvira Gonzalez de Mejia, "Caffeine (1, 3, 7-Trimethylxanthine) in Foods: A Comprehensive Review on Consumption, Functionality, Safety, and Regulatory Matters," *Journal of Food Science* 75, no. 3 (2010): R77–87.

28. Peter J. Rogers, "Caffeine and Alertness: In Defense of Withdrawal Reversal," *Journal of Caffeine Research* 4, no. 1 (2014): 3–8.

29. B. M. van Gelder et al., "Coffee Consumption Is Inversely Associated with Cognitive Decline in Elderly European Men: The Fine Study," *European Journal of Clinical Nutrition* 61, no. 2 (2007): 226–32; and Eduardo Salazar-Martinez et al., "Coffee Consumption and Risk for Type 2 Diabetes Mellitus," *Annals of Internal Medicine* 140, no. 1 (2004): 1–8.

30. Emma Childs and Harriet de Wit, "Subjective, Behavioral, and Physiological Effects of Acute Caffeine in Light, Nondependent Caffeine Users" [in English], *Psychopharmacology (Berlin)* 185, no. 4 (2006): 514–23.

31. H. A. Young and D. Benton, "Caffeine Can Decrease Subjective Energy Depending on the Vehicle with Which It Is Consumed and When It Is Measured" [in English], *Psychopharmacology (Berlin)* 228, no. 2 (2013): 243–54.

32. Paracelcus, as cited by Wikiquote from "Die Dritte Defension Wegen Des Schreibens Der Neuen Rezepte," *Septem Defensiones* 1538, vol. 2, (Darmstadt, 1965): 510. English translation via Google Translate, http://de.wikiquote.org/wiki/Paracelsus.

33. Astrid Nehlig, "Is Caffeine a Cognitive Enhancer?" *Journal of Alzheimer's Disease* 20, suppl. 1 (2010): S85–94.

34. Karen Alsene et al., "Association Between A2a Receptor Gene Polymorphisms and Caffeine-Induced Anxiety" [in English], *Neuropsychopharmacology* 28, no. 9 (2003): 1694–702.

35. Tad T. Brunyé et al., "Caffeine Modulates Attention Network Function," *Brain and Cognition* 72, no. 2 (2010): 181–88.

36. A. S. Attwood, S. Higgs, and P. Terry, "Differential Responsiveness to Caffeine and Perceived Effects of Caffeine in Moderate and High Regular Caffeine Consumers" [in English], *Psychopharmacology (Berlin)* 190, no. 4 (2007): 469–77.

37. Peter J. Rogers et al., "Faster but Not Smarter: Effects of Caffeine and Caffeine Withdrawal on Alertness and Performance" [in English], *Psychopharmacology (Berlin)* 226, no. 2 (2013): 229–40.

38. Meagan A. Howard and Cecile A. Marczinski, "Acute Effects of a Glucose Energy Drink on Behavioral Control," *Experimental and Clinical Psychopharmacology* 18, no. 6 (2010): 553–61.

39. David A. Camfield et al., "Acute Effects of Tea Constituents L-Theanine, Caffeine, and Epigallocatechin Gallate on Cognitive Function and Mood: A Systematic Review and Meta-Analysis," *Nutrition Reviews* 72, no. 8 (2014): 507–22.

STRATEGY 5: MAKE YOUR WORKSPACE
WORK FOR YOU

1. James L. Szalma and Peter A. Hancock, "Noise Effects on Human Performance: A Meta-Analytic Synthesis," *Psychological Bulletin* 137, no. 4 (2011): 682–707.

2. Juliane Kämpfe, Peter Sedlmeier, and Frank Renkewitz, "The Impact of Background Music on Adult Listeners: A Meta-Analysis," *Psychology of Music* 39, no. 4 (2010): 424–48.

3. Göran B. W. Söderlund et al., "The Effects of Background White Noise on Memory Performance in Inattentive School Children," *Behavioral and Brain Functions* 6, no. 1 (2010): 4.

4. Gianna Cassidy and Raymond A. R. MacDonald, "The Effect of Background Music and Background Noise on the Task Performance of Introverts and Extraverts," *Psychology of Music* 35, no. 3 (2007): 517–37.

5. Patrik Sörqvist and Jerker Rönnberg, "Individual Differences in Distractibility: An Update and a Model," *PsyCh Journal* 3, no. 1 (2014): 42–57.

6. "Loudness Comparison Chart (dBA)," South Redding 6-Lane Project, California Department of Transportation, www.dot.ca.gov/dist2/projects/sixer/loud.pdf.

7. "Loudness Comparison Chart (dBA)," http://www.dot.ca.gov/dist2/projects/sixer/loud.pdf.

8. Ravi Mehta, Rui (Juliet) Zhu, and Amar Cheema, "Is Noise

Always Bad? Exploring the Effects of Ambient Noise on Creative Cognition," *Journal of Consumer Research* 39, no. 4 (2012): 784–99.

9. David M. Berson, Felice A. Dunn, and Motoharu Takao, "Phototransduction by Retinal Ganglion Cells That Set the Circadian Clock," *Science* 295, no. 5557 (2002): 1070–73; and S. Hattar et al., "Melanopsin-Containing Retinal Ganglion Cells: Architecture, Projections, and Intrinsic Photosensitivity," *Science* 295, no. 5557 (2002): 1065–70.

10. Russell G. Foster, "Neurobiology: Bright Blue Times," *Nature* 433, no. 7027 (2005): 698–99; and David M. Berson, "Phototransduction in Ganglion-Cell Photoreceptors" [in English], *Pflügers Archiv: European Journal of Physiology* 454, no. 5 (2007): 849–55.

11. Antoine U. Viola et al., "Blue-Enriched White Light in the Workplace Improves Self-Reported Alertness, Performance and Sleep Quality" [in English], *Scandinavian Journal of Work, Environment, and Health* 34, no. 4 (2008): 297–306. Philips Lighting, which makes the lightbulbs used in the study, supported the research.

12. F. Ferlazzo et al., "Effects of New Light Sources on Task Switching and Mental Rotation Performance," *Journal of Environmental Psychology* 39 (2014): 92–100.

13. K. C. H. J. Smolders, Y. A. W. de Kort, and S. M. van den Berg, "Daytime Light Exposure and Feelings of Vitality: Results of a Field Study During Regular Weekdays," *Journal of Environmental Psychology* 36 (2013): 270–79.

14. Anna Steidle and Lioba Werth, "Freedom from Constraints: Darkness and Dim Illumination Promote Creativity," *Journal of Environmental Psychology* 35 (2013): 67–80.

15. Robert Desimone and John Duncan, "Neural Mechanisms of Selective Visual Attention," *Annual Review of Neuroscience* 18 (1995): 193–222; and Stephanie McMains and Sabine Kastner,

"Interactions of Top-Down and Bottom-Up Mechanisms in Human Visual Cortex," *Journal of Neuroscience* 31, no. 2 (2011): 587–97.

16. Gilles Pourtois, Antonio Schettino, and Patrik Vuilleumier, "Brain Mechanisms for Emotional Influences on Perception and Attention: What Is Magic and What Is Not," *Biological Psychology* 92, no. 3 (2013): 492–512.

17. Thomas W. Malone, "How Do People Organize Their Desks?: Implications for the Design of Office Information Systems," *ACM Transactions on Office Information Systems* 1, no. 1 (1983): 99–112.

18. Dana R. Carney, Amy J. C. Cuddy, and Andy J. Yap, "Power Posing: Brief Nonverbal Displays Affect Neuroendocrine Levels and Risk Tolerance," *Psychological Science* 21, no. 10 (2010): 1363–68.

19. Josh Davis and Pranjal H. Mehta, "An Ideal Hormone Profile for Leadership, and How to Attain It," *NeuroLeadership Journal* 5 (forthcoming).

20. Andy J. Yap et al., "The Ergonomics of Dishonesty: The Effect of Incidental Posture on Stealing, Cheating, and Traffic Violations," *Psychological Science* 24, no. 11 (2013): 2281–89.

21. Florence-Emilie Kinnafick and Cecilie Thøgersen-Ntoumani, "The Effect of the Physical Environment and Levels of Activity on Affective States," *Journal of Environmental Psychology* 38 (2014): 241–51.

22. Marily Oppezzo and Daniel L. Schwartz, "Give Your Ideas Some Legs: The Positive Effect of Walking on Creative Thinking," *Journal of Experimental Psychology: Learning, Memory, and Cognition* 40, no. 4 (2014): 1142–52.

23. Ruth K. Raanaas et al., "Benefits of Indoor Plants on Attention Capacity in an Office Setting," *Journal of Environmental Psychology* 31, no. 1 (2011): 99–105; Eleanor Ratcliffe, Birgitta Gatersleben, and Paul T. Sowden, "Bird Sounds and Their

Contributions to Perceived Attention Restoration and Stress Recovery," *Journal of Environmental Psychology* 36 (2013): 221–28; Mathew White et al., "Blue Space: The Importance of Water for Preference, Affect, and Restorativeness Ratings of Natural and Built Scenes," *Journal of Environmental Psychology* 30, no. 4 (2010): 482–93; and Gregory A. Laurence, Yitzhak Fried, and Linda H. Slowik, "'My Space': A Moderated Mediation Model of the Effect of Architectural and Experienced Privacy and Workspace Personalization on Emotional Exhaustion at Work," *Journal of Environmental Psychology* 36 (2013): 144–52.

INDEX

ABOUT THE AUTHOR

Josh Davis, Ph.D., is deeply connected to both the scientific and business communities. He received his bachelor's from Brown University and his doctorate from Columbia University. His academic work concerns the topics of embodied cognition, emotion regulation, and the neuroscience of emotion. He has taught at Columbia University, New York University, and Barnard College of Columbia University. At Barnard, Josh was a fulltime faculty member in the department of psychology for five years, prior to transitioning to his current role. In business, Josh began his career in mechanical engineering before his doctoral work. He now serves as the director of research and lead professor for the NeuroLeadership Institute. In this role, he guides the Institute's work in translating basic science research for business and leadership use, and drives strategy for new research content. Josh also coaches and trains individuals in the areas of public speaking and productivity. He lectures frequently both for NeuroLeadership Institute events and at conferences and organizational events worldwide. Josh's published work in the business realm includes blog posts or articles in *Harvard Business Review, strategy + business, People & Strategy, Training + Development, Psychology Today,* and the *NeuroLeadership Journal* and has been featured on *Harvard Business Review*'s "The Shortlist." Josh grew up in Alexandria, Virginia. He lives with his wife, Daniela, in New York City.